Tales of the West Somerset

Recollections of some of those who helped bring the line back to life

Edited by Ian Coleby,
Allan Stanistreet and Ian Tabrett

Published by the
West Somerset Steam Railway Trust

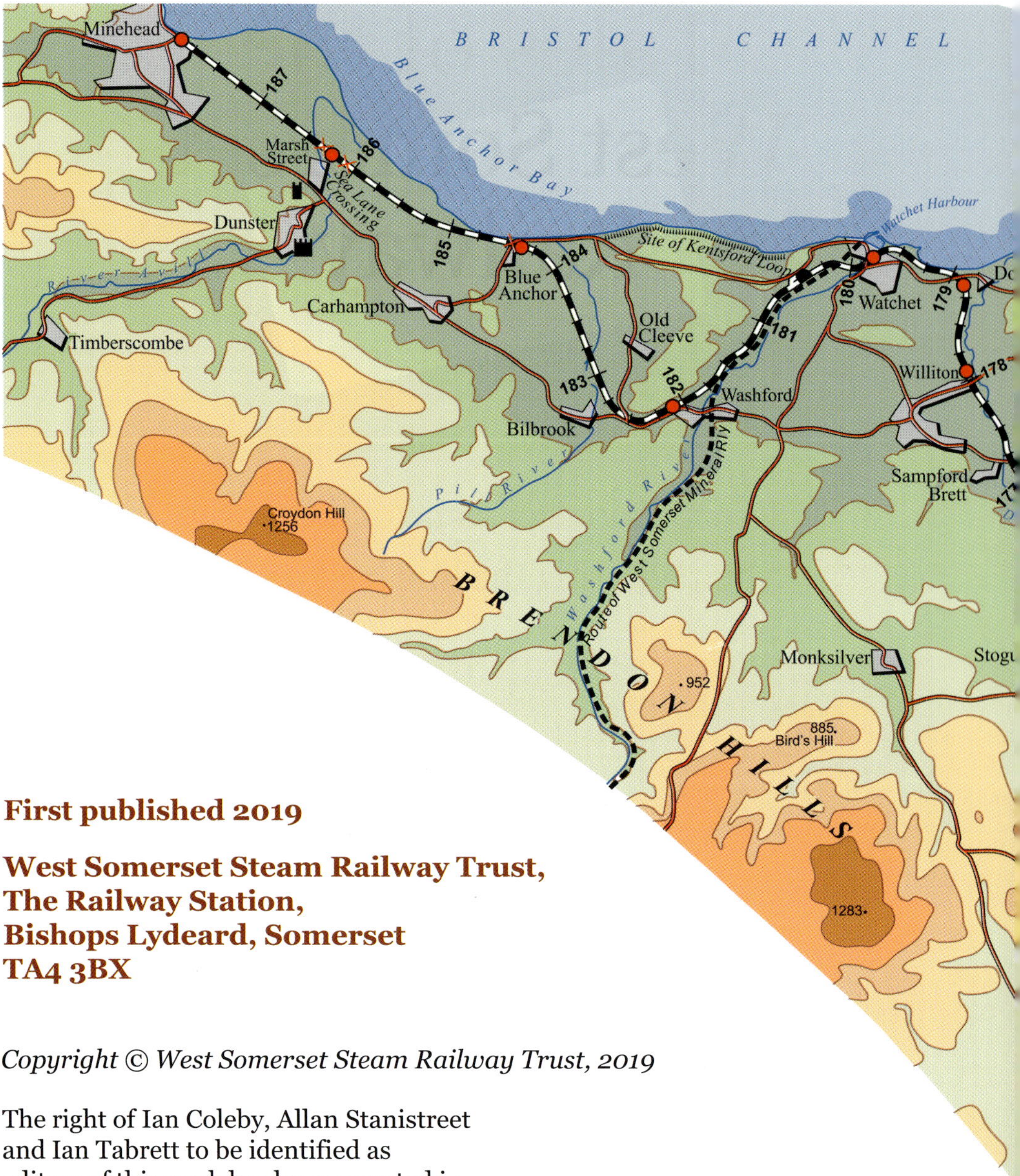

First published 2019

**West Somerset Steam Railway Trust,
The Railway Station,
Bishops Lydeard, Somerset
TA4 3BX**

ISBN 978-1-5272-3877-0

British Library Cataloguing-in-Publication Data.

A catalogue record for this book is available from the British Library.

Printed in the United Kingdom, 2019

Printed by Short Run Press, Exeter.

THE WEST SOMERSET RAILWAY

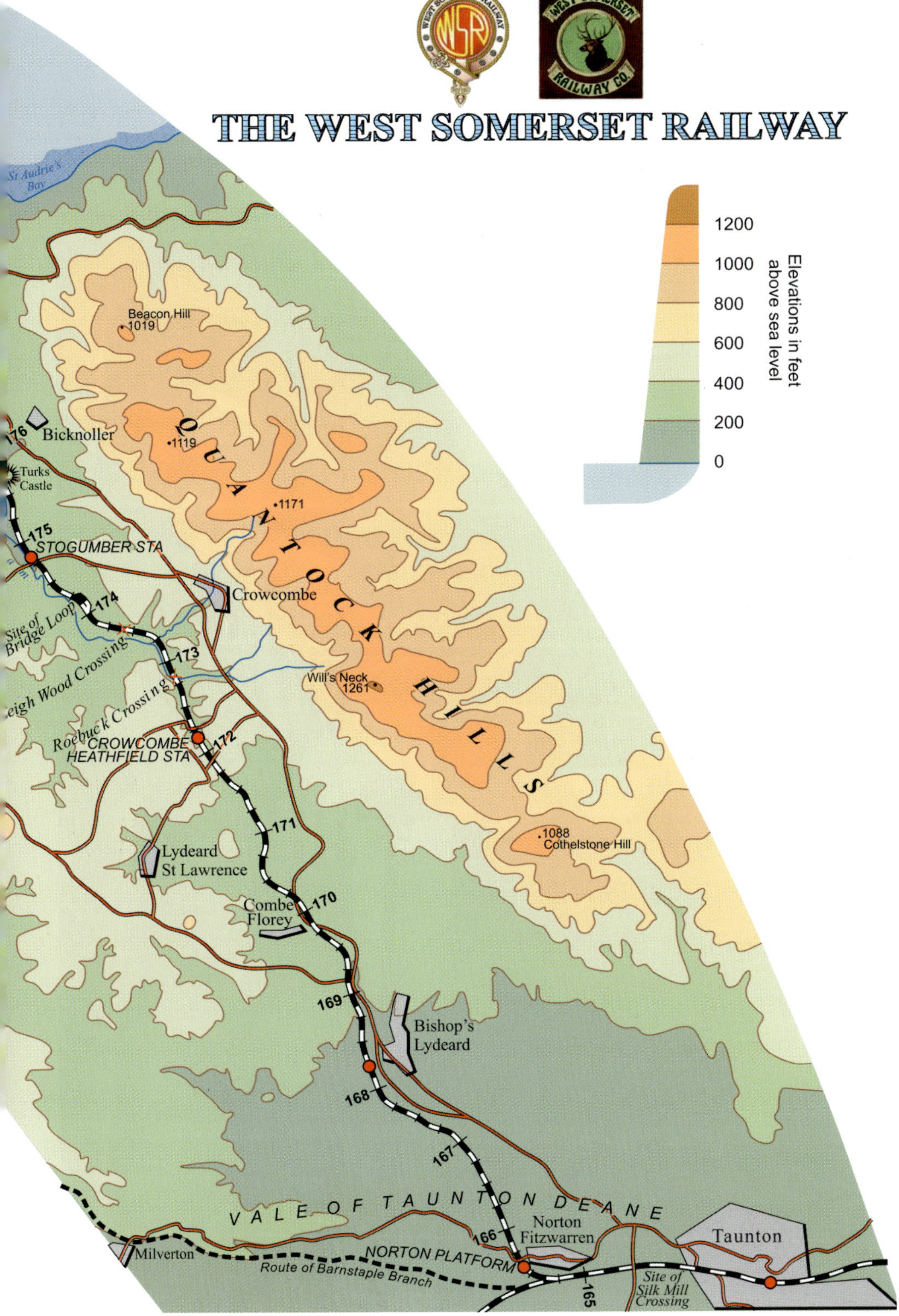

Gil Bird (left) and Arthur Young on the footplate of 2994 Vulcan. Allan Stanistreet

Contents

Introduction

This book has been produced to celebrate the 40th anniversary on June 9th 2019 of the West Somerset Railway running trains from Minehead to Bishop Lydeard. It celebrates that magnificent achievement by recalling the stories and memories of just some of the huge band of people who made that possible.

We should not underestimate the effort that was required to reopen the longest standard gauge heritage railway in the country and the many hours of difficult work to achieve that. We owe all of them a huge debt of gratitude for what they have achieved and it is our hope that this book goes just a little way to saying thank you by recognising their efforts.

The book has been an absolute delight to put together and has given the editors their own trip down memory lane in reading these memories and trawling through photographic collections to illustrate the book.

We are pleased that the West Somerset Steam Railway Trust has taken on the publication of the book as part of their remit to document the heritage and history of the railway.

All profits from this book will go to support the railway and all the editors' and contributors' time has been given free. We hope you enjoy and appreciate it as much as we do.

The editors would like to dedicate the book to all the railway people who worked so hard on the line and who are no longer with us.

Ian Coleby
Alan Stanistreet
Ian Tabrett

Foreword

I was elected in a by-election in March 1970. During the campaign there had been rumours of a threat to the Taunton-Minehead railway, but these were firmly denied. When I took my seat in Parliament, the very first Ministerial letter I received was from Mr Fred Mulley, the Minister of Transport, to tell me that it had been decided not to pay the subsidy for the line, and it would therefore close. Edward Du Cann, the MP for Taunton, and I tried hard to get the Minister to change his decision, but without success. The future looked bleak, but then a great fight-back started, led by the redoubtable Douglas Fear and a splendid supporting cast. I gave all the help that I could, but the overwhelming credit goes to them, and it is their achievement that we commemorate on this 40th anniversary. How much satisfaction they have got out of all their hard work, and how much enjoyment they have given to the countless numbers of their passengers over the years as they travel through the lovely countryside and relive the thrill of steam.

One of my happiest memories in one election campaign was when I did a genuine whistle-stop tour on the railway from Watchet to Minehead, mimicking the old American routine!

I am delighted that there is this book to commemorate the first 40 years, and I wish it every success.

Lord King of Bridgwater
March 2019

Acknowledgements

The editors gratefully acknowledge the contributions of all of those who have taken time out of their railway lives to write articles for this book. We owe all of these contributors and the many other volunteers and paid staff a great debt for the countless hours they have spent working on the railway, often very long and unsocial hours.

We say a great thank you to the photographers featured in this book: the late Peter Barnfield, Mike Dunse, Steve Edge, Walter Harris, John Jenkins, Mike Johns, Nick Jones, Tom King, Claire Sheppy, Allan Stanistreet, Peter Triggs, Alan Turner, Robin White, the late Tony Yardley.

We are very grateful to Nigel Power who has worked hard as technical photo editor to prepare the many pictures for reproduction. Thank you.

Thank you to the West Somerset Steam Railway Trustees who have supported our work and to Jonathan Jones-Pratt who gave the initial impetus to get us going!

Minehead on re-opening day, 28th March 1976. Victor *is about to leave for Blue Anchor with the first train. A triumph for all who worked so hard to get to this point.* Allan Stanistreet

How the Line was Lost and Regained

Lord King of Bridgwater

This article was written by Lord King, then MP for Bridgwater, and was published in the local press in 1976. It is reprinted here with some small amendments with the permission of Lord King.

The very first letter I received from a Minister came two days after taking my seat in the House of Commons after the by-election in March, 1970. It was from Mr Fred Mulley, who was then Minister of Transport, advising me of his decision to cease paying a grant for the Taunton to Minehead railway line and accepting the British Rail proposal to close it.

This came as a great shock as there had been no mention of this whatsoever during the by-election, and there were certainly many who felt it was hardly a coincidence that the announcement did not appear until after the by-election had taken place. Whatever the truth of that, there was no doubting the seriousness of the decision since it meant that West Somerset would lose a key lifeline in an area where communications were already inadequate. British Rail had been asking for a subsidy of £140,000 a year to operate the line in the form of grant aid from the Minister, and it was the refusal to agree to this grant that in turn gave consent to the closure by British Rail. There was considerable surprise at the size of the subsidy that BR were requesting, and there has since been much criticism in the House of Commons of the BR accounting system for branch lines. Fortunately, the general concern about the closure of the line was widespread, and a small group of enthusiasts led by Mr Douglas Fear set about seeing if there was any way in which the line could be reopened as a private railway company.

A series of extremely complicated negotiations commenced between the West Somerset Railway Company and BR to establish terms on which BR would dispose of all the land in their possession and the track representing the Taunton-Minehead line. Somerset County Council became interested, partly from an interest in the railway land around the station at Minehead and also because of the obvious social benefit in restoring the passenger service from Minehead to Taunton. There were questions raised at the time as to the prudence of the county council purchase of this property but I always felt that it was a perfectly sound investment. Not only did they acquire valuable land in Minehead which would not be required for the actual railway operation but also the scrap value of the track itself is considerable as an ultimate security if the railway operation totally failed to succeed.

Once the purchase had been completed it was necessary for the legal formalities to be completed for the transfer to the county council, and the issue of a Light Railway Order. Railway law is an extremely complex matter and anyone who has read the enormous detail of the final orders which specify all the rights related to every level crossing, footpath, right-of-way and so on will appreciate how complicated it is.

This is an all-too-brief summary of the many meetings, negotiations and discussions that have taken place to bring us from the original announcement of the closure in 1971 to the welcome prospect of the reopening of the line now in 1976. This has involved an enormous amount of work by Mr Fear and his colleagues who deserve every congratulation on the way in which they have kept their objectives in spite of all the frustrations and difficulties that have arisen. I have tried to help in whatever ways I can, particularly in trying to remove delays and unnecessary blockages, whether at the Department of the Environment or British Rail, but there is no question that the real credit goes to Mr Fear and his colleagues.

And so now the railway company has reached a point where the tourist steam excursions will shortly be starting from Minehead to Blue Anchor, and it is hoped that before too long it will be possible to start again the service down the whole line from Minehead to Taunton station. A great help in this has undoubtedly been the money received under the job-creation programme which has enabled many people to be employed to speed up the clearance and restoration of the line. The real question will therefore soon be answered as to whether the private railway company can operate the line profitably where British Rail incurred such losses. Many have expressed doubts about this but I have much in mind a remark once made to me by Sir Richard Marsh, chairman of British Rail. He said on one occasion that the two most profitable parts of BR were InterCity and steam excursions. I'm not sure whether we can call Taunton to Minehead InterCity! However it is certainly a very useful passenger link and before its closure was carrying a significant number of passengers on the early morning run from Minehead, and the evening return.

But the line is going to develop the steam excursion activity and there is no doubt that this is an enormous attraction. Moreover, it has a priceless advantage of linking to the mainline railway station at Taunton and I understand there are already discussions on the possibility of BR running trains from other parts of the country right through to Minehead. This should also be of benefit for visitors to Butlins; it also has the additional interesting feature that it runs immediately beside the docks at Watchet, and there is therefore a further possibility of developing a freight traffic with the docks. These factors then do suggest there are real prospects that the railway company can be a success, but a key factor must be whether people in West Somerset will support this enterprising venture.

Recalling the Early Days of the WSR

David Butcher

In the autumn of 1975, I was approached to consider becoming the first General Manager of the West Somerset Railway. Knowing only a little about the Minehead branch, I carried out a simple survey to assess the prospects. The main concern was whether the track rail-depth remaining was sufficient to last for at least a few years, because of the high costs of replacement. Discreet enquiries to former BR friends revealed a most useful diagram with details of each track length and of the planned renewals envisaged. This, together with the site survey to check surviving facilities, allowed a suitable judgement to emerge: against some offered advice to decline, I decided to accept the appointment and on 1st January 1976, the task began.

One of the earliest objectives was to recruit a secretary, and the very welcome arrival of Kath Grainger was achieved. She quickly became a very respected member of the team and did all sorts of non-secretarial jobs as well. Very soon a gang of up to 60 Government-sponsored unemployed workers was recruited to tackle the considerable arrears of maintenance that had accrued since final closure five years earlier. It became a real challenge to manage this together with the increasing number of Association volunteers becoming available to undertake specific requirements. We were also able to appoint a very small number of paid staff to oversee the key functions of track (Harold Blackmore), motive power (Harry Lee), and operating (Nick Jones), supplemented in due course by steam fireman and diesel unit driver (Roy Hartle), guard (Don Spencer), signalman (Jim Elson), driver (David Rouse) and track (John Sumbler), plus my wife Linda and her small team in that most important area – the buffet, ready for when the passenger trains began running from 28th March 1976. Considerable volunteer help covered other key functions, particularly the Signal and Telegraph contribution led by John Jenkins, who installed the Dunster Sea Lane automatic level crossing. John is still offering his services on the line but sadly most of these early pioneers have since taken their leave of us.

The effort required to re-open to Blue Anchor successfully surprised our Railway Inspecting Officer, Major Peter Olver, who had doubted our ability to achieve it in so short a time, and we were suitably congratulated. Major Olver wasn't so surprised when asked to inspect the next section on to Williton in the August, and we re-opened to there on the 28th in time for the very welcome Bank Holiday traffic. We had planned this to be by mid-July to secure the full holiday business, but other events prevented this.

Meanwhile the Board had authorised the training of the footplate staff to work over BR into Taunton, and in August this was undertaken. Harry Lee, David Rouse and myself became qualified as steam and DMU drivers, as we all possessed extensive previous main-line footplate experience, and Roy Hartle became qualified as a DMU driver and steam engine fireman. Unfortunately, soon afterwards other events arose to prevent us working into Taunton, but it does prove that from the earliest days the plans to do so were being developed.

The idea had been to achieve the intended re-opening through to Taunton on Saturday 4th December 1976, with an intermediate opening to Bishops Lydeard in October. By June the line was fully cleared of all undergrowth through to Norton Fitzwarren, and all drainage had been cleared and restored along the line. Our Government work-experience employees had achieved this with an average of between 30 and 40 men, the full number being rarely available.

Interior of Blue Anchor signal box on opening day. From left, Norman Solomon, Jim Elson and General Manager, David Butcher. Allan Stanistreet

The remaining work and significant expense required were for the two level crossings at Leigh Wood and Roebuck to be converted to automatic operation, to enable through working to be implemented. Unfortunately, these became delayed, although an intermediate re-opening to Stogumber was introduced in 1978 capable of being operated only by the DMUs due to the absence of a run-round facility there for locomotives. I believe a very rare steam operation did work to Stogumber, but with an engine marshalled at each end of the train.

We were fortunate that all station buildings had been boarded-up by British Rail immediately after the line was closed. This meant these had survived better than expected, although still requiring much work to become suitable again for public use. At Minehead, we inherited a very basic track layout with just the main and bay platforms, each of which had separate run-round loops. Both of these had been worked from Dunster with two single lines to Minehead where no signalling provision or signal box remained. At Blue Anchor, a nearly-complete signalling system had been re-instated by the volunteer S & T team but it was not yet in proper working order. At Williton, remarkably the original Bristol and Exeter brick-built signal box had survived to become the sole one of this type anywhere, although all the signalling and associated instruments had been dismantled by BR. At Crowcombe, the signalling and the signal box had gone entirely, with just a single line remaining in the up platform. At Bishops Lydeard, all signalling and equipment had gone with just the empty signal box surviving. There was a lot of work to be done!

Apart from the Dunster Sea Lane crossing, the one other major requirement was to provide basic maintenance capability for the locomotives and diesel units at Minehead. Major Olver was prepared to stretch a point regarding the need for a track inspection pit but required a minimum provision of one track for motive power requirements. We also needed to create a run-round in the main platform as to rely on the bay platform would have depleted the passenger potential because it could not be seen from the seafront roadway. Harold Blackmore, with volunteer help, created both the run-round and the siding ready for March 28th. It was an incredible success story to equal the superb progress being made by the S & T staff, with so many notable efforts made to reach the re-opening date. The S & T work at Blue Anchor and Dunster and the Minehead track alterations, were absolutely critical.

Earlier, before the Light Railway Order had been granted, our two steam engines, *Victor* and *Vulcan* plus the two DMUs and other stock had been stored in the old Taunton engine shed area to enable a great deal of preparatory work to be carried out, led by Harry Lee and his volunteer team. Finally, a week before opening, the GWR Pannier Tank No. 6412 was purchased by the Association from the then Dart Valley Railway to improve our motive power requirements. All the diverse events coming together provided a very special moment when the first train hauled by *Victor,* suitably adorned with decorative bunting, began its historic journey after the opening speeches had been made.

Few who witnessed it will ever forget it. And then to see so much effort for the line to be re-opened to Bishops Lydeard by 1979 is another huge achievement to be celebrated now, 40 years on.

The two Bagnall 0-6-0 locomotives came from South Wales via Birmingham. Although modern design, they were shunting engines and totally unsuited to hauling passenger trains over the West Somerset. Nevertheless, they kept the railway supplied with steam traction in the earlier years. Above, we see 2994 Vulcan *newly arrived at Minehead and below, 2996* Victor *ready for its next journey along the line.*
Both: Allan Stanistreet

Early Days at Minehead Loco

Frank Houlding

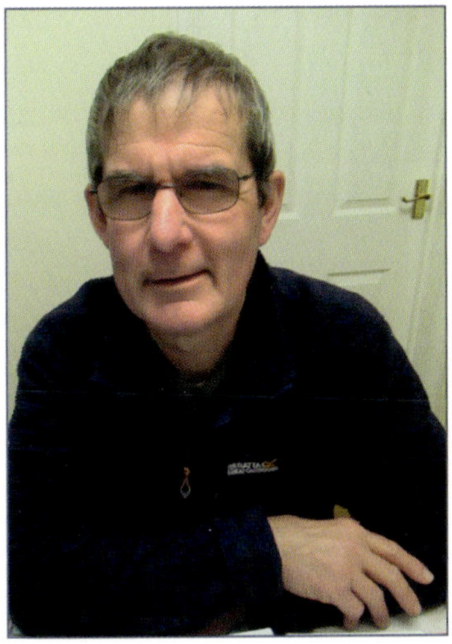

Looking at the many almost state-of-the-art facilities around the WSR today, it is easy to forget, if you ever knew, exactly what existed when a few folk decided that the line must not be lost. Take the current motive power depot at Minehead, for example: workshops of truly impressive proportions, tools of recent manufacture, plenty of machinery and techno capability, as well as a well-fitted running shed and servicing facilities that make today's train crews' lives so much easier.

In the early 1970s the goods shed – that is what the nucleus of the current loco shed was in pre-WSR days – when we first ventured in was derelict and somewhat sinister, with a strange aroma that you just couldn't or would rather not, put your finger on. It was clear from the general bustle and shuffling overhead that there was a major infestation of pigeons. There were gigantic (without a lie, 12ft in many cases) stalagmites of pigeon poo that were thickly distributed around the interior. This was the start of a long-running battle with these airborne residents. Every trick in the book was tried over many years to try and keep them out.

The whole place was attacked with shovels, masks, and extremely bad language as it had to be levelled, tidied and de-brambled before the shed could be filled with our sparse stock of loco spares and equipment which initially consisted of a few items from the late, great Harry Lee's garage which was a gold mine of old Great Western surplus. We had to be ready for the day when we had engines, or even some track to lay into the shed. As opening day grew closer the place even boasted a kettle in the office, some shelves in the store room, a working toilet and a sink to wash in. Once trains were running regularly it became clearer exactly what sort of life could be expected in the fledgling locomotive department. In order to coal the locos it was necessary to barrow from the coal pile, which would be at ground level wherever the incoming tipper driver had left it. Then up the ramp of what had been the loading dock and tip it on the deck at footplate level. Once the deck was full there was often a pause to catch breath before getting stuck into the pile with a shovel, throwing the coal up over the top of the bunker side-sheets until you were drained and it was full. It was just as well that the bunkers of *Victor*, *Vulcan* and No. 6412 held only two tons of coal, as it all had to be moved at least twice, meaning four tons of work!

After the prep, if no-one had turned up to fire (a few characters would turn up magically just as the bunker was full) you had the job and off you went. Admittedly, it was not an arduous fire to Blue Anchor and back in the very early days, but there was lots to be learned. The end of the working day was really not for the faint-hearted as the fire had to be dropped on those days that our mentor had decided the fire needn't be cleaned for two or three days, something we never understood. But the final nail in a tough day's work was to crawl under the rear of the loco to hose and scrape the ashpan clean and try to flush it all out with, if it was working, the "good" hose, around two inch bore, somewhere up near what is now the shed throat hand-point. It rapidly became a game of "where is there a gap in the clinker" to flush the loco without having to shovel a space clear to avoid being coated in yesterday's mess of ash and spoil as you crawled through.

Another problem among many that emerged in our initial year of operation was the hellishly hot summer. Some days the fire brigade would be chasing around our lineside fires, while on others they would arrive in the nick of time to fill an empty tank engine in desperate straits. Sometimes we would even set off towards Blue Anchor with a petrol-driven pump, a strainer and a great helping of optimism to find a trickle of water in a brook to ensure that operations could continue!

For my own part, in the early 1970s, I had finished my education and was now working

Opening day, 28th March 1976 sees pannier tank 6412 and Victor *being prepared by the engine shed staff.* Peter Triggs

The original rather crude facilities have improved in this more modern photo which shows the engine shed extension. SDJR 7F No 88 simmers outside while long-standing volunteer and staff member Stuart Nelhams is seen inside the shed. Allan Stanistreet

my way along the lower levels of the British Rail footplate world, having been really lucky to have been hired at Taunton. Neville Bond and I were the last two to be hired while the diesel-hydraulics were still in service, and a better depot there never was. Most turns involved stepping on to an express passenger, and mostly we got to just step off one again at the end of shift, Taunton being the mid-way crew-change point for most services. All the other depots on the Western Region hated us for it! Driving was almost guaranteed for half the shift, hopefully with a helpful and friendly experienced mate as we rarely got the same driver twice in a week. Of course, there were the occasional down sides, like the spare turns where you were likely to have to try and rescue some failing 12-coach train with Taunton's resident pair of Class 31 diesel-electrics which Bath Road had banished our way once they found that they would not work at all unless coupled in multiple together because of some exotic electric fault. Even together they could hardly make it up over Whiteball.

It was a very proud day when I drove a Paddington-Plymouth express into Taunton and my driver (Lionel Cox, a great mate) dived across the cab, right across my seat and vision, and joyfully thrust his head out of the slider window yelling: "'Ere Frank, there's a ruddy steam engine in the Loco!" Thankfully I managed to stop on the mark and was keen to get a look at it too. Once the relief crew arrived it was a case of us both going to inspect the new arrivals. *Victor* and *Vulcan* were being explored, prodded and poked by a sizeable number of Taunton loco staff. Up to that point my tales of what was about to happen on the WSR had been all pipe dreams to them, but now there was something to show that we meant business.

At the time I often worked eight hours for BR, eight hours for the WSR, and then grabbed what was left to sleep in the back of my van in the Taunton station car park. A hard core of around 20 would gather and be thrown at the most pressing problem of the moment. Meanwhile there was a chilling realisation that the new HST sets that had been on trial were about to take over most of the mainline express work, and faster timing meant crew changes at Exeter. For us at Taunton that meant likely redundancy on a "last in first out" basis, so I began to panic and ended up with my latest of many attempts at getting a driver-vacancy transfer being accepted...to Inverness in just two weeks! Hearing this, Harry Lee, then CME of the WSR and long-time mentor, friend and Barry scrapyard rescue captain, stepped in and offered me a DMU driver/steam fireman post.

The die was now cast and to the sound of Taunton driver Ray Harrison pleading "Don't do it, Frank, we need to keep you keen young 'uns," I left Taunton loco for the Minehead shed. There will never be a day that I don't miss my time on BR on the Class 1000s, and though I did grow to really like the Class 50s, once the axe had fallen on all the "old girls," it was never going to be the same again. A whole new world opened up, and with my old friend Roy Hartle being the other DMU driver, many an enjoyable day was had, shock though it was in the winter to have to take turn as guard every other week. Roy was one of the WSR's unsung heroes: a devoted servant of the company who did lots of work with a dry wit and tenacity. He could turn his hands to many things, and we were intrigued and not a little excited one bright autumn morning to hear of his latest "special purchase": a shower! For several days he toiled with the assistance of a volunteer electrician to hitch it up to the power supply but with the rapidly-lowering temperatures as the season went on, we had to be really keen to strip off in the gale that often howled through the shed to make use of it.

Such were the everyday trials and triumphs of West Somerset Railway loco life back then, and it is a constant amazement to me just how far the line has progressed over the years, leaving me with the satisfaction of knowing I made steam driver back in the 90s as a plus, but with expensive osteopath bills on the minus side!

Harry Lee waits for the "right away" while driving Victor.
Tony Yardley

How a Boy's Dream of Engine Driving Became Reality

The story of Harry Lee, reprinted from an article published in the local press in 1976.

Harry Lee's lifelong love affair with steam all started with a steamroller when he was six. Today, he's still in steam, taking trippers from Minehead to Blue Anchor and back on the West Somerset Railway, and loving every minute of it. At 52, life for Harry has become "one long paid holiday," to use his own expression. He was born near Torquay, the son of a police constable who took his son's interest in engines seriously. His excitement at seeing a steamroller on the road outside the garden gate led to dad having a word with the driver and, in no time at all, Harry was up there among the polished levers and pressure gauges discovering a fondness for steam which has endured. "I couldn't leave it alone after that first ride," he said.

When he was nine his father put his name down with the GWR, for those were the days when entry into such a calling was highly competitive. Meanwhile he haunted the engine sheds, and from the age of 10 until he was 13 was frequently on the footplate, helping the firemen stoke up on the old Class 4555 engines on the Lydford to Launceston and Tavistock run: "Quite unofficially, of course. That sort of thing simply couldn't happen today."

Harry Lee oiling round at Bishops Lydeard. Tony Yardley

His family moved to Barnstaple where there was a loco shed with five engines. Then out of the blue came a letter on Great Western Railway official notepaper inviting Harry, who had just turned 15, to apply for the vacancy as a cleaner. That job might have been the lowest rung of the railway ladder, but in those days they didn't employ just anybody. He polished his boots, put on his best suit and travelled by train (not on the footplate this time) for an interview with Mr James, Locomotive Foreman at Taunton.

Mr James was inclined to approve, but that was not the end of it. A fortnight later came another train journey, this time to Swindon – that Mecca of all railway aspirations – where Harry underwent exhaustive medical tests along with eight other applicants. Only two were successful, one of the lucky ones being Harry Lee. A fortnight later he started his job as a cleaner in the Barnstaple shed, part of the great GWR empire, in which he learned to call even the chargehands "Sir". As for such luminaries as Mr. James at Taunton, Harry remembers that they walked around resplendent in bowler hats and frock coats.

Every locomotive and all the rolling stock had to be turned out not only functionally sound but also visually immaculate. After only three weeks of this Harry had a lucky break when a regular fireman fell ill and the beginner was instructed to act as his replacement. He took to the work like a duck to water and everyone could see that Harry was in his element. Within a year he was sent to Westbury in Wiltshire to become, at the age of only 17, the youngest fireman on the whole of the GWR.

It was hard work, but harder still because of the turmoil of the war years. Conditions were difficult, manpower short and the hours long. Harry recalls that on one occasion he worked for 42 hours without a break. Everything was in short supply and eventually it became difficult to find enough coal to keep the locomotives rolling. Even house coal was resorted to and it flared away at a furious rate in the fireboxes. Many a time, on the one-and-a-quarter hour run from Westbury to Taunton, Harry single-handedly shovelled seven tons of house coal out of a tender into the firebox.

It was back-breaking work compared with the sort of things he sometimes does now in his capacity as locomotive superintendent of the West Somerset Railway, where he drives the diesel unit with no more effort than to press a button and turn a lever. But although he is proud of the company's diesel unit, his heart is still in the lovely old steam engines which are painted up as pretty as pictures and are in use on the line. "Steam engines – why they're human," he said, "diesel isn't, it responds to electricity and oil." To fully understand what he means, you have to take into account his lifelong love affair with steam engines together with all the heartbreak of the post-war decline in railways and the countless hours of voluntary work he has put in more recently restoring old engines from the scrapheap.

During the war he had been sent to Barnstaple and later Taunton with the duties of a passed fireman, which is a railway manner of saying a trainee driver. That way he obtained many hours' experience driving on branch and main lines. Some of the lustre went out of life when diesels began to replace steam at the end of the 1940s, and 10 years later Harry had left the railway and was running his own, a scale-model track at Clevedon with an engine that would pull a load of 60 adults. He spent five years building and running model locomotives, one of which he sold for £600 that later changed hands at £7,000 and went to the USA.

Of Harry Lee's association with the Great Western Society and subsequently with the West Somerset Railway, of his tireless efforts spending all his weekends in every holiday over at the Barry scrapyard rescuing old locomotives, he simply says: "A whole book could be written." He is intensely proud of the West Somerset Railway, the longest privately-owned line in the country, and is unstinting in the praise he accords to all the workers and volunteers who have made and are continuing to make miracles happen so that the line can operate successfully.

On March 28th, when he left his home at Eastwick Road, Taunton, and went to Minehead to drive the steam train which officially opened the West Somerset Railway, it was perhaps the happiest day of his life, comparable only with that magic moment when at the age of six he was lifted up for his first ride on a steam roller!

The Rush to Re-Open, 1976

Nick Jones

On 1st March 1976, having left my job as a signalman at Witham and with express trains, locals and stone traffic now in my past, I started work as the WSR's first Operating Superintendent. I'd been approached by WSR General Manager David Butcher, who was recruiting experienced railway staff to re-open the Minehead branch in stages and run an intended DMU year-round public transport service. We'd be restoring a railway which had lain virtually unused since closure by British Railways early in January 1971. Notable exceptions were the storage of long strings of unwanted goods wagons awaiting a trip to the scrapyard, and the movement of ex-LMS 4-6-2 express locomotive 6229 *Duchess of Hamilton* from display at Butlins, Minehead, to Taunton and on to overhaul at Swindon. The WSR had first gained full access to the line in December 1975, and had run a Directors' Special from Bishops Lydeard to Minehead and back just before Christmas, but the line was in no fit state to run passenger trains.

Some work had already got underway, but not much. The West Somerset Railway Co Ltd had very limited resources. There was huge local support for the re-opening: the people of the district wanted their train service back. The West Somerset Railway Association had been formed to support the effort to reopen the branch and to provide a focus for the many people who wanted to offer voluntary support. Weeks before my starting date, David Butcher, Locomotive Superintendent Harry Lee and Permanent Way Superintendent Harold Blackmore had started work. On the day I arrived, a group of 35 temporary workers also started on the WSR, employed as part of a Government Job-Creation Programme. They had been recruited on a six-month contract to provide the labour that would fight back against the takeover of the line by nature – clearing vegetation from the lineside, repairing and repainting buildings, assisting with track work, fencing, and drainage. Voluntary effort had also started, with work going on at several sites to repair track and signalling, to work on motive power and rolling stock such as it was. There was an air of industry about the place, fortified by a lot of interest

from press and tv, and local well-wishers.

The plan was to re-open the entire line from Minehead right into the bay at the country end of the Up Relief platform at Taunton as quickly as possible. The first section would be from the seaside terminus to Blue Anchor, some three-and-a-half miles. This would involve a number of tasks, not least the conversion of the two level crossings at Dunster so that they would comply with the new Light Railway Order. Dunster West would be an open crossing with a 5mph speed restriction. Sea Lane, on the Taunton side of the station, would become an open crossing with road traffic controlled by lights and local monitoring of operation by drivers' white light signals, subject to a 10mph restriction. The order for the level crossing equipment had been placed already. Along with John Jenkins, who worked in the S&T Department for BR at Gloucester and had taken a lead in the work needed to make the WSR's signalling fit for the re-opening, I wasn't happy to discover that although the road lights and the train drivers' white light signals were all on order, no thought had been given to how a train would trigger the operation of the crossing.

The directors had decided the first section would re-open on 28th March, just four weeks from my starting date. The railway would be running trains over the Easter holiday and would be able to attract educational parties from among the thousands of schoolchildren staying at Butlins in their five-week programme of geography field-trips for schools. The race was well and truly on. To add to the pressurised environment, the directors had also decided that the company would go public on re-opening day with the launch of a public share issue with a minimum subscription of 50 shares (£5) in an effort to restore their railway service. We could not be seen to fail – the railway had to be ready in time for a pre-opening inspection by the Railways Inspectorate, due on 22nd March. It was going to be an uphill run.

How little did the railway actually possess by way of tools and equipment, locomotives and rolling stock! The Job-Creation Programme allowed the purchase of a few basics such as shovels and scythes. The locomotive department owned nothing except what Harry Lee had collected in past years – useful stuff but not much with which to run a railway. The Permanent Way department had a few iron bars and key hammers that had been scavenged. The Signal & Telegraph Department functioned on whatever the volunteers brought with them. Locomotive-wise there was 0-6-0ST *Victor,* a fleet of one. And *Victor* didn't yet have the means of vacuum-braking a train. The WSRA had two ex-BR Mark I coaches, neither of which had a brake compartment. The WSR had two 2-car Park Royal DMU sets which had been parked in sidings at Taunton since they had been purchased from BR on the transfer of the Paignton to Kingswear service to the Dart Valley Railway. The list of jobs went on and on. Some would be completed and ticked off only to be replaced by other newly-discovered tasks.

The only way the re-opening could be achieved was by massive extra effort from the relatively small number of volunteers working with the far smaller team of us who were on the payroll. It was going to be close. The PW Department were busy reinstating the country end of the run-round loop at Minehead, which had been removed by BR. This was in addition to fettling the whole length of track to Blue Anchor, fencing, cutting back vegetation, installing a lead and siding off the run-round loop at Minehead to provide access to the goods shed, which was to become the loco shed. Electricity supplies for Dunster and Blue Anchor had to be laid on, plus a suitable water supply at Minehead to keep the locomotives running.

The S&T Department had the level crossing at Sea Lane, Dunster, to install and commission as well as find, purchase and install a two-lever ground frame at the Taunton end of the run-round loop at Minehead. The level crossing at Blue Anchor and the pointwork, signals, and lever frame had to be got back into use. During the closure the large wheel in the signal box by which the crossing gates were swung had been removed but fortunately it was donated back by its owner and had to be re-instated and commissioned.

At a meeting with David Butcher a week before the inspection by the Railways Inspectorate was due, he provided me with a list of jobs to be completed, including getting the upgrade of Sea Lane Crossing for which we were still awaiting the late-ordered components. To say we were getting a bit twitchy would be an understatement.

Opening day, 28th March 1976 sees Victor *adorned with the Union Flag about to attach to the first train to Blue Anchor.* Peter Triggs

A lot was riding on this. There was still no vacuum-braked locomotive and precious few seats to convey our hoped-for passengers. This could so easily have gone very wrong. At a very late stage negotiations for the WSRA to purchase ex-GWR 0-6-0PT locomotive No. 6412, which was surplus to requirements on the Dart Valley and Torbay railways, were concluded. The WSR Company still didn't have funds to make the purchase so it fell to the volunteer organisation. At the same time, four ex-BR Mk I coaches came up for tender at York and these were somehow purchased by the WSR Company. They were brought by freight service to the WSR via Severn Tunnel Junction. New volunteer Chris van den Arend boarded the coaches there to ride with them in a freight transfer to Taunton.

Time was so short that we could not afford a slip-up. The four coaches comprised a BSO, a brake vehicle; an SK corridor compartment coach; an SO open coach converted from a first class vehicle, which had the grubbiest imaginable seats; and a TSO open coach similar to the two already owned by the WSRA. With these on the line, we had a train suitable to run the intended service.

As things turned out these four coaches, No. 6412, and a hired auto trailer having steamed up from Paignton with BR crew, arrived at Taunton in time to be taken to the boundary between BR and the WSR Milepost 165¼ at Norton Fitzwarren ready to be collected by *Victor* on the morning of Thursday 25th March, just four days before the re-opening.

Work on the signalling, track, fencing, signage, and level crossings at Dunster had been completed ready for inspection by Major Peter Olver of the Railway Inspectorate, on Monday 22nd March. I look back now on this scramble to have the line ready in so few days with astonishment. It's not as if there were scores of people. Only very few were involved and it is an indication of their dedication that the target date was met.

The sequence to light up the road signals at Dunster's Sea Lane crossing to stop road traffic was initiated by treadles and occupation of track circuits. The long spell of disuse had led to a build-up of rust on the rail heads, preventing operation of the track circuits. This we discovered just the day before the inspection so it was a case of getting hold of a drill with a few abrasive disks and polishing off the rust and running a train of camping coaches, hauled by *Victor* over the crossing and back several times to shine the rails.

Despite a number of shortcomings requiring attention before the reopening day, Major Olver trusted us sufficiently to say that he would recommend the Secretary of State to approve the line for passenger use. Phew! We had made it! One of the tasks was a significant one: *Victor* still didn't have a vacuum brake. The necessary ejector arrived a couple of days before the reopening, and the team in the Loco Department worked day and night to get it fitted and working. They eventually finished in the small hours of Sunday 28th March with the flagging-off of the first train by Lord Montagu of Beaulieu only hours away.

It brought a lump to my throat, to raise my arm to signal the right away to Don Spencer, guard of the first train. He duly waved the green flag to Harry Lee and Cedric Dunmall on the footplate of *Victor*. With the deep chime of the loco's whistle, the six coaches rolled out of Minehead's platform watched by a huge, cheering crowd. We were off, but it was a close-run thing.

45 years with the Signal and Telegraph

John Jenkins

In 1972 I started on the British Railways Signal Engineering Scheme based at Gloucester Depot. I'd had a keen interest in railways for as long as I can remember and had decided to make it my career. The scheme involved day-release for college and various elements of district work, from maintaining, altering, and installing signalling equipment to working on the automatic telephone exchanges and railway telephone network. Contrary to general perception, the S&T section is an entirely engineering department while the signalmen are part of the Operating Department, using the equipment but not undertaking any work on it.

In the spring and summer of 1973 I was attached to the Worcester signal gang, who were installing new semaphore signals and point-operating equipment at the Worcester boxes in what was to be the last major mechanical re-signalling scheme undertaken by the Western Region. It is nearly all still there being more intensely worked than ever before.

I had a special affection for the recently-closed Minehead branch, having used it for family holidays since the late 1950s, and I was keen to see it re-opened as soon as possible. I realised that my newly-gained knowledge and skills would be of some use and to that end I wrote to the WSR in August 1973 to offer my assistance to whoever was to bring the signalling equipment back to use. To my surprise the reply from Major Tony Griffiths, the Company's interim Manager, was that there was nobody yet in place to do that, and inviting me to Taunton to discuss it. I then had a few days to visit all the crossings and stations with Major Griffiths, and it became clear that despite my age and limited experience, the Company would like me to lead the work necessary. So that autumn I became the first S&T volunteer!

There was a great deal involved. At the end of BR's operations, there were only signal boxes at Williton, Blue Anchor and Dunster, with resident crossing keepers at Roebuck, Leigh Wood and Sea Lane, Dunster. All these sites were substantially complete in respect of the mechanical apparatus but all electrical equipment and the line wires from the telegraph pole route had been removed soon after closure, which had become standard to prevent theft – particularly in the case of dead-end branches. Copper wire secured premium prices.

The new Company's plans meant the employment of as few staff as possible, and it was agreed with the Department of Transport's Railway Inspectorate (now ORR) that Dunster signal box would not reopen and that the level crossing there and at the three residential crossings would be converted to Automatic Open Crossings with flashing lights. Dunster West crossing had such little traffic that it could simply be open with road Give Way signs. The S&T work therefore included designing and installing the new crossing equipment, restoring the mechanical signalling to operational condition, and providing the minimum of electrical repeaters and controls at the signal boxes. Use was to be made of GPO telephones, with single wooden train staffs for each single-line section. The line still belonged to BR at this stage and no work by the WSR was permitted, so my initial activities were for me to become conversant with Automatic Open Crossings, of which few existed at that time, and to start planning the equipment which would have to be purchased.

In the summer of 1974, BR agreed to the WSR having limited access to carry out superficial work to prevent further deterioration of the track and signalling equipment. This was interpreted as getting as much as possible back into working order without actually putting back anything that was missing! With a few other young chaps, the first place we tackled was Blue Anchor where we started to get the points working again from the signal box. The Taunton-end points were 350 yards away and we had to clear masses of vegetation even to see the rodding and cranks, which were all seized solid. The normal way to free cranks at that time was to disconnect the rodding, apply copious quantities of oil, put a lot of cotton waste around and set fire to it! When they were hot, the crank arms would be hammered and generally they would eventually move – sometimes needing several lots of treatment. There are still cranks in use today bearing the hammer marks applied all those years ago.

This work was repeated many times over the next 18 months until eventually all the points were operational. For the signals, we recovered wires from the various distant signals which were no longer to be operational, and this gave us enough to allow various stop signals and point detection equipment to become operational by the end of 1975.

By then, the WSR had full control of the line and the pace of work increased. On the first weekend of 1976 we replaced the gate-wheel at Blue Anchor which had been sold by BR in 1971 – fortunately to a WSR supporter who kindly returned it – although a spare had already been acquired from BR when the crossing at Awre Junction was converted to barriers back in December 1973. With the gate-wheel back in place we recoupled the gates, and despite gale force winds we managed to get them across the road and back across the railway. However our delight soon turned to woe when we found that one of the rising back-stops in the middle of the road wouldn't drop back down as it was seized in the raised position, standing some 6ins high in the middle of the road. No amount of hammering would make it go down but thankfully road traffic was light and we managed to direct traffic out of the way. One vehicle caught us unawares and hit the stop but this had the desired effect and it went back down,

Work on Minehead groundframe. Jim Almond (left), and Jon Francis, aged 15 and 22 respectively at time of photo. Jon went on to become President in 2006/7 of the Institution of Railway Signal Engineers (IRSE). John Jenkins

seemingly with no damage to the vehicle. More oil was poured down and when tried again a week or so later, it worked correctly.

For the line to be opened from Minehead to Blue Anchor on 28th March 1976, Sea Lane level crossing had to be completely altered. The existing large gates and their posts had to be removed and flashing lights and all the control equipment had to be purchased and installed. I'd already produced designs for it all and once Major Peter Olver, of the Railway Inspectorate, had agreed the final layout and light positions we could press ahead. The closure of the former Midland station at Gloucester the previous December conveniently took out a number of lineside equipment cupboards. Douglas Fear, the Company's first Chairman, came to Gloucester with a van to collect a load of these and other useful equipment which BR had sold to me for £10. By mid-March all the equipment for the crossings had been bought and I was released by BR and paid by the company to lead the installation with a good number of volunteers.

One of the first things to do was to run a cable back to the station to provide power for the crossing because all the signalling equipment in BR days was run off primary batteries with no mains supply whatsoever. I had already decided that the WSR would not use primary cells because they are a constant expense and work load. We were then faced with the next headache: the crossing was operated by track circuits which needed relatively clean railheads. With no regular use for more than five years, the rails were very rusty and we had to clean them, the only way being to run a number of trains.

Victor, with a few motley coaches and goods vehicles, ran over the crossing about 50 times to clean enough rust to make the track circuits work. With a few days to go, Major Olver passed the crossing and the rest of the line from Minehead to Blue Anchor as fit and compliant, so all was set for the grand reopening on the following Sunday.

With the line now open, more volunteers came to the fore and we could turn our attention elsewhere. At Watchet, the former BR warning bell for the footpath crossing was deemed inappropriate by HMRI and we were required to install automatic red/green lights because most of the projected rail traffic would be DMUs running at up to 40mph. The installation was carried out in the sweltering heat of 1976's summer. Like Sea Lane we had to install track circuits, and drilling the rails by hand for the bonding round the rail joints over almost half-a-mile took some stamina. At Williton, nearly all the signals were invisible to the signalman so electrical indicators and track circuits had to be installed as well, although most track circuit bonding had been left by BR so could be used. We also had to install lineside cables to connect it all to the signal box because the old telegraph pole route was not going to be reconstructed.

Like most of his colleagues, Major Olver was a retired Army officer and he had joined HMRI in 1967. He was a good friend to the WSR and had been a Somerset cricket supporter since his father was Rector at Luccombe in the 1940s. He generally timed his visits to us when they were playing at Taunton and he could combine the two things.

Sea Lane Crossing was one of the biggest jobs undertaken by the S&T for the initial re-opening in March 1976. Seen here, the new crossing lights have been installed and the old GWR wooden gates, now redundant, lie discarded. John Jenkins

He was very down-to-earth and once declared to council Highways Officers that he wasn't a typical Civil Servant when giving his very practical interpretation of some obscure requirement. He told them: "Where it is stated 'the Minister shall decide,' that means me – so watch out!"

Another interesting job was the removal of Dunster signal box in November 1977. HMRI required us to move it to improve visibility at the now open crossing and it seemed sensible to move it to Minehead where one would eventually be needed. The move by rail was like a military operation, precisely planned and led by Hugh Perrett, and No.1163 Whitehead did the honours at walking pace.

As well as maintaining what we had in use, we had to acquire what mechanical signalling we could as it was removed from the main line. One of the biggest of these jobs was the entire removal of the 99-lever locking frame from Westbury South in October 1978, which gave us enough to install the frames for Minehead and Bishops Lydeard boxes years later. Next we installed two more automatic crossings in the Quantocks for the opening to Bishops Lydeard in June 1979. The biggest task was laying a power cable for the mile-and-a-half from Crowcombe Station to Roebuck and Leigh Wood Crossings because there was no mains power anywhere in the vicinity.

My work with BR had, from late 1976, taken me to the Signalling Design Office at Reading, and it was now time for another move. In October 1981 I was appointed to be the S&T Line Manager for the North and West line between Newport and Shrewsbury. This was mainly mechanical signalling so I put a lot of the knowledge I'd acquired from WSR into practice for BR! The down side of this meant I could no longer come to the WSR regularly but kept in touch by phone and did some equipment restoration away from the railway. This situation, together with a young family, continued until the late-1990s, although on my appointment to Bristol we moved to Somerset. By now the WSR had expanded considerably from the early days, with three more signal boxes and much bigger locomotives.

In 2003 Mark Smith asked if I would be prepared "on a short-term basis" to be the "Professional Head" for Signal Engineering. A year later he said it seemed to be working so could it become permanent? I couldn't resist and my still-voluntary role is to act as advisor to the Board and to provide technical direction on S&T activities. The department is still principally an unpaid band of loyal individuals supporting one full-time and one part-time employee. For all of them I have the utmost admiration since most are past 65 and carry out a vast number of tasks, from routine maintenance and the few failures to erecting signals and level crossing equipment as well as a lot of electrical signalling installation.

Three major schemes have been finally achieved. The biggest has been the return to the traditional Electric Token working for the whole line, giving back that flexibility in working which the original WSR Train Staff and Ticket never could. It was no mean feat as quite apart from the overhaul of many instruments, the burying of nearly 23 miles of cable was undertaken, one of the penalties for being the longest operational standard gauge heritage line. Another was the re-commissioning of the main line connection at Norton Fitzwarren to full passenger standards. As a result, we now have WSR-owned and maintained signalling interface equipment with the Exeter network control area. This prestigious scheme has given us considerable flexibility for through working including NR's Ballast Cleaner train. More recently we have completely re-signalled Williton, giving far more operational flexibility for today's operations. This has involved relocking the existing lever frame, renewing all electrical signalling equipment with full track-circuiting of all running lines, and installing several new signals.

The biggest challenge for the WSR S&T now is to keep the signalling equipment safe and reliable. Much of it was second-hand 30-plus years ago and it is wearing out faster than we are able to replace it with our voluntary staff resources. The other thing that has changed is the far greater degree of external regulation we now have, principally due to the plethora of standards and regulation necessary when BR was broken up but applying almost as much to the heritage railways connecting with the national network.

Many amusing incidents occurred in the early years, most of which are unprintable although one that can be revealed concerns the volunteer accommodation at Minehead. At one time the mess and sleeping coach was stabled in the Bay Siding. It had double doors which opened from the dormitory area to the loading dock. One individual only worked on Saturdays and in the evenings he drank copious quantities of beer and would sleep on until about midday, defying all attempts to wake him earlier. He snored very loudly so one balmy night the others decided that for them to sleep, he needed to go elsewhere. Complete with his bunk he was taken through the double-doors on to the platform under the canopy and to their amazement he was still there slumbering when passengers for the mid-morning trains were filing past...like some kind of "lying in state"!

Peckett 1163 Whitehead *had been bought by members of the GW Society and acted as Minehead station pilot. The locomotive left the railway in 1982. It is seen here at Minehead in 1977 coupled to 6412 wearing* Flockton Flyer *nameplates.* John Jenkins

Early Work on the Stations and Buildings

Brian Stainton

My first experience with the West Somerset Railway was in July 1975 when I was working in the printing works at Williton. During my lunch breaks I used to wander around to the station hoping that one day it would reopen, not knowing that meetings were being held to reinstate the line from Minehead to Taunton. On one occasion I saw Bill Thrush, who lived a few doors away, on the track-bed pulling weeds, and I took up his invitation to help during my spare time. The main priority was to open Williton Station. Bill and I worked to reinstate the men's lavatory. The pipes had been severed so I used some copper pipe fittings I had to repair it. We struggled to find the stop cock, which we eventually discovered near the overgrown signal box and we were able to get the lavatory working again. However, having cleared the pan of debris and flushing it, the water stayed there. The sewer pipe from the station to the road near the crossing gates was full of silt, as were the inspection chambers. However, I dug down and cleared all the debris so that finally, joy of joys, the water flowed again and the lavatory was functioning.

Bill then introduced me to the West Somerset Railway Association and at the same time my exploits at Williton led to me being asked to help with the plumbing arrangements at Crowcombe Station. I was happy to do so, and the railway eventually had working facilities at both these stations. Work was now proceeding at Minehead with a view to opening the line to Blue Anchor, where I became involved with – surprise, surprise – the gentlemen's loo, which was in a state of disrepair. Once again I replaced the pipes and got it working, although at that time it was open to the elements as there was no roof.

Incidentally, I am not a trained plumber but my DIY skills were very welcome at this time. I was also helping with cutting-back at Blue Anchor and Crowcombe Stations, when it was realised ridge tiles and slates at Blue Anchor were missing and water was leaking in; I managed to acquire the necessary ridge tiles by hook or by crook as there were no funds available at the time, and I fitted them to prevent further water damage. When it was discovered the signal box had broken and missing slates, up I went again, although it was considerably higher than the station roof.

On Sunday 28th March 1976 the great day arrived and the first train from Minehead to Blue Anchor set forth pulled by *Victor* with great pomp and circumstance. The tape was cut by Lord Montagu and the band played with great gusto. People were all over the track with their cameras, myself included. As soon as the first train departed, I was engaged cleaning the carriages for the second train of the day, which I was privileged to ride on at a cost of 50p. I still am the proud possessor of the ticket.

I remember assisting S&T at Blue Anchor when it started to snow and we retreated to the signal box, lit the fire and had a jolly time enjoying the wonderful view from on high, as no trains were running at the time. I was living at West Quantoxhead and it snowed so much that we were eventually cut off. But before it got too deep, a frantic phone call came from Williton Station to say a pipe had burst in the roof and water was cascading everywhere. No one knew where the stopcock was, so down I went, turned the water off and in due course after the thaw, carried out the repair.

My work continued at Crowcombe, mainly cutting-back. When the company approached me to suggest I might like to open up Stogumber station, I was delighted. Part of the platform was causing concern because much consisted of old sleepers that

Early work at Williton involved much weeding of the track bed. Here a volunteer gang is seen doing just that with pickaxes and shovels. Backbreaking work in the hot summer of 1976. Allan Stanistreet

had rotted away and were in a dangerous condition. I was joined by a husband and wife who ran a sub-post office in Taunton and who brought a chain saw to help dismantle the structure. He cut up the sleepers and sold them as kindling in his shop to raise money for Stogumber. Joe Raucki, a builder based in Bishops Lydeard, lent us his dumper truck to help with the work. We built a ramp from surplus ballast to make the platform accessible and safe, reducing the length by a third. Tom Waller turned up to work on the garden wearing a kilt; he lived in a caravan in the grounds of the Railway Hotel. Once again, I found myself on the roof replacing ridges and slates but as there was no water at this time, thankfully I had no work involving lavatories!

I had now been elected on to the WSRA committee as Stations and Buildings Officer and was invited by the company to open Watchet Station, which was boarded up. When I finally got into the main doors, I cautiously walked in to find a hole in the floor. The whole station was in chaos, the ladies' lavatory was broken and the floorboards were rotten. The glazing bars in the window frames were broken and the window glass non-existent. I acquired glazing bar off-cuts from a local business, and window glass which I fitted as best I could. Others popped in to help when they could, including Hector Davidson, who discovered an entrance at the platform end of the building that had been bricked up. The door was in working order so he knocked the brickwork down, opening up the station where the shop is now. Hector also took a liking to the former goods office, taking possession of it and using it to store his tools. Allan Stanistreet aptly named this "Hector's House."

Local youngsters enjoyed removing the newly-replaced window putty, so it was a case of two steps forward, one step back. I caught them at it one day, and showed them how to use the putty to fix the windows, so giving them a job that was well done by them and the windows are there to this day. Appeals went out for fund-raising; sponsored walks were organised on sections of the line and newspapers were amassed and stored in Bishops Lydeard goods shed until the Watchet Paper Mill collected them and paid us well. Jumble sales were held, and in fact any event we could think of to raise money.

Walter Harris was concerned about the booking office chimney at Crowcombe which was curving towards the booking office roof and in imminent danger of collapsing. After consultation, it was decided I should reduce the height of the chimney to a safe level.

In October 1978 I was sent by my employer to Bournemouth University on a course for 12 months, and when I came home on the Friday evening of the first week, there was a plea from Hugh Perrett because help was needed at Williton to reinstate a DMU engine. I arrived at about 6 o'clock and my task was to connect the fuel pipes to the injectors plus other ancillary bits while the rest of the gang reinstalled the starter motor. Came the moment to put the engine back into the power car there wasn't enough space between the carriage and the workshop wall! Result: four men had to put their backs to the end of the power car to push it out into the open where there was sufficient room to reinstall the engine and connect fuel pipes, cooling pipes and transmission to our satisfaction. We finally fired it up at about five the next morning.

The first Santa train was the DMU, with the guard's compartment converted to a grotto. Guess who drew the short straw to be Santa? It was good fun, though, especially when two young ladies plied me with a glass of wine!

The point leading into the goods siding at Blue Anchor was lifted and moved to Minehead to become the lead into the engine shed. These two photos shows that hard, manual work in progress with nothing more than hand tools. The lower photograph appears to show a sound recordist on the left hand side and there is a thought that he may have been from HTV. Both: Allan Stanistreet

Volunteering on the WSR in the Early Years 1976-78

Tim King

I was a teenager at Yeovil in the 1960s and went to boarding school in Wells during the time of branch-line and station closures. My earliest railway memories included watching the Bulleid Pacifics go through Yeovil Junction, the Westerns and Warships on the West of England main line at Bruton and an 08 shunter waiting at the level crossing in Wells next to the Sherston Hotel. We used to walk along the redundant Yeovil to Taunton line at Montacute and explore the old lines around Hendford in Yeovil.

In 1968 our family holiday was on the Somerset coast at the Luttrell Estate's caravan park at Blue Anchor. My favourite occupation was to stand and wave at the blue BR diesel units as they slowed down for the station and, heading in the up direction, the drivers changed gear for the assault on the bank to Washford. My parents were persuaded to let us go on a train trip to Minehead, my first visit to what seemed like a vast and mainly disused seaside terminus.

In the early 1970s I became involved in my first railway preservation project, the Somerset & Dorset Railway Circle's scheme to create a steam attraction at Radstock station. My duties included applying copious amounts of red lead paint to the frames and other parts of loco No. 53808 which had been rescued from Barry scrapyard. David Shepherd opened his new base at Cranmore in 1973, and I was a steward on the opening day of the East Somerset Railway.

In September 1974 I went to Sussex University at Brighton, a long train journey from Bath Spa. For a few years I had been following developments on the West Somerset Railway through reports in the *Western Daily Press* and regional TV news programmes.

What attracted me was the proposal for a proper railway with trains for local people between Taunton and Minehead, with a seasonal steam service along the coastal section. I applied to join the West Somerset Railway Association, at a cost of 75p, and became member number 240. My earliest correspondence with Membership Secretary

George Hitchcock and the Company in October 1974 survives, together with my Association Newsletters from that time.

Company Secretary Mrs Audrey Fear wrote to me on 16th October 1974, including the following paragraph: "Our proposal is to have a daily diesel service between Taunton and Minehead, running about six trains a day in each direction. We have British Rail's agreement to lease the line from Norton Fitzwarren to Taunton Station with a bay platform there. During the summer we intend to have steam trains running frequently between Minehead and Williton as a holiday attraction and, if required, through trains from Taunton. We are also hoping that it will be possible to deal with freight to and from Watchet harbour."

I bought shares in the first issue and kept up to date with developments via the monthly newsletter. Eventually I made it to Minehead in April 1976 for my first stint at volunteering. I arrived by bus and reported to Kath Grainger, the General Manager's secretary, who pointed me in the direction of Nick Jones, the Operating Superintendent. Both were very welcoming and I was given a smart red coat with red deer logo and red WSR tie and told to go and find Don Spencer, a BR veteran with long service at Dunster and Minehead from the late 1940s. He took me under his wing and I learned how to check and sell tickets and perform the duties of a guard, which included cleaning out the coaches at the end of the day then washing the outsides so they looked presentable to the travelling public.

As it was Easter the trains were busy and there was still the sense of excitement at the

Early volunteers were faced with endless amounts of backbreaking work to clear the track of the accumulated weeds and brambles. Here the gang is seen working at Dunster. Allan Stanistreet

beginning of a brand-new venture. The newsletters had conveyed a little of the sometimes-fractious relationship between the Company and the Association. However, everyone at Minehead and along the line to Blue Anchor got on very well and pulled together as a team. As a 19-year old joining a new organisation where I didn't know anyone was a little daunting to begin with but I soon settled in and made friends. We slept in a camping coach parked next to the bay platform. There were mess facilities and a shower in the old goods shed, which housed the Loco Department. The locos in traffic were the Bagnall *Victor* and the GWR pannier tank No. 6412. The Company also had two Park Royal DMUs from the Kingswear line. In the bay sidings could be seen my old friend Somerset & Dorset loco No. 53808 and Hymek D7017.

From General Manager David Butcher, to drivers, guards and signalmen there was a wealth of railway experience among the staff and volunteers. It was a pleasure to hear their stories when we had a quiet few moments over a cuppa in the buffet or the mess room. Locomotive superintendent Harry Lee lived in a second camping coach and would sometimes invite the volunteers in for a chat and to talk about the old days of steam, including his time on the footplate of General Eisenhower's private train during the war.

On Sunday evenings a big vat of sausage stew might be cooked by Linda Butcher and her catering team. Inevitably a large amount of time and money was spent after work in Minehead's numerous hostelries. This was a great time to make newcomers to the team welcome and to cement our friendships. Some in my case have been life-long.

Later in 1976 the line was extended to Williton, which opened up a whole new world for the train crews and enabled passengers to visit the working port of Watchet. It also provided a new challenge for the steam locos with Washford Bank to be tackled on a daily basis. I spent as much of my holiday time and weekends at Minehead as my studies would allow. It was a bit of a trek by train from Brighton to Taunton, with draughty waits at Westbury and watching the clock in the hope that the 19.55 bus to Minehead wasn't missed! In June 1976 I left university never to return and spent the rest of the year on the railway. I worked as a Guard on the early morning DMU services to Williton as we started to build up a group of regular passengers – ladies from Washford heading for the Van Heusen shirt factory at Watchet, and commuters and shoppers going to Minehead.

However, these services were not

WEST SOMERSET RAILWAY Minehead 4996

We're back in business! A new Diesel service

STARTING SATURDAY, DECEMBER 11th, 1976

WINTER SHOPPERS' SPECIALS

		Tuesdays & Fridays only			Saturdays only		
MINEHEAD	dep.		12.30	16.30		12.30	16.30
Dunster	arr.		12.35	16.35		12.35	16.35
Blue Anchor	arr.		12.42	16.42		12.42	16.42
Washford	arr.		12.49	16.49		12.49	16.49
Watchet	arr.		12.56	16.56		12.56	16.56
Williton	arr.		—	—		13.01	17.01
Williton	dep.	—	—	—	09.55	13.55	17.15
Watchet	arr.	10.01	14.01	17.21	10.01	14.01	17.21
Washford	arr.	10.08	14.08	17.28	10.08	14.08	17.28
Blue Anchor	arr.	10.15	14.15	17.35	10.15	14.15	17.35
Dunster	arr.	10.21	14.21	17.41	10.21	14.21	17.41
MINEHEAD	arr.	10.27	14.27	17.47	10.27	14.27	17.47

We regret that owing to engineering works there will be no service to Williton on Tuesdays and Fridays. No service on Christmas Day.

SPECIAL DAY RETURN FARES from MINEHEAD

		Children & O.A.P.s
Dunster	25p return	15p return
Blue Anchor	40p "	25p "
Washford	55p "	35p "
Watchet	65p "	45p "
Williton	65p "	45p "

Timetable subject to alteration without notice

its your railway USE IT!

to last as they were too costly to run. In December I started a new job as a barman in the Beach Hotel opposite Minehead station where I had my own room at the back and worked four days a week for £25, all meals included. This meant I could carry on working on the railway, but I missed out on a starring role in the TV series *The Flockton Flyer* as I had to work one of the days when the Indian Mutiny scene was shot.

One of the delights in the early years were trips to recover bits of unwanted railway gear. A gang of us set off one early morning with a lorry and a van to Dai Woodham's scrapyard at Barry. It was still full of redundant steam locos and we spent many hours climbing over Bulleid Pacifics and various Great Western classes. A few parts were purchased for the three prairie tanks the Association had bought and removed to Bishop's Lydeard in 1975. We also went for a day trip to the Longmoor Military Railway in Hampshire and brought back all kinds of kit including the wheelset that still sits under the Minehead station sign on the platform.

TV filming was a regular occurrence in those years with *The Flockton Flyer*, *The House of Eliott*, *Shoestring* and *The Lion, The Witch and The Wardrobe* to name just four. My big break came in 1978 with fellow volunteer Tim Maddocks when we were selected to be in an episode of *Some Mothers Do 'Ave 'Em* with Michael Crawford as Frank Spencer. We were requested to be on duty from early morning and eventually got to film our scenes around 4pm. I was on the platform with Michael and Tim got to wave his green flag as the guard. It was a fun day!

I joined the Association Committee as Publicity Officer in 1977 and edited the first WSR Stock Book. In 1988 I returned to Minehead as the first (and last) Tourism Officer at West Somerset District Council. But that is another story.

Early volunteering on the track. A trolley fresh from more clearance is pushed into Blue Anchor. Steve Martin is apparently doing the hard work! Allan Stanistreet

My Life and Times on the West Somerset Railway

Steve Martin

I am a born-and-bred Minehead resident, my father was brought up in Quay Street and mother was from Watchet and later lived in Washford. She came from a GWR family and was employed as a porter at Minehead during the 1940s. As children we were regular visitors to the station and we used the railway a lot to visit relations or for trips to Taunton. Our last was the week before closure, a little saddened but having no idea of its future impact on my life.

News of the scheme to reopen the line re-awakened my interest in railways, and in July 1972 when I was walking past Minehead Station I saw a caravan advertising the preservation scheme. I was invited in by the late John Pearce who was running the display. We discussed the project and he extracted £2.50 from me to join the West Somerset Railway Association. Regular newsletters appeared giving details of progress on the scheme and that the attempt to purchase the line had been overtaken by Somerset County Council's decision to buy the line to safeguard it for the future.

Then came the announcement of working parties at Taunton Station on the railway's Fruit C Van, so one evening I travelled there, wandered down the platform past a Western loco that was just starting up to this railway van. John Wood and Mike Grimoldby were working on it, and this was the start of many days of working together and of lasting friendships. I was asked to join the committee of the WSRA, a position I held for five or six years, meeting at the Black Horse pub in Taunton. As preparation of legal arrangements for transferring the railway and a Light Railway Order moved slowly on, the Association was allowed access to stations to clear weeds and carry out other work.

Graham Harwood in his Morris van painted up in GWR colours oversaw Saturday working parties while Mike Grimoldby, in his blue J4 van, oversaw Sundays. We cleared every station, involving a lot of digging out weeds from ballast and some memorable bonfires. One day at Minehead somebody leaned over the railings and suggested: "You're wasting your time." I remember volunteers such as John and Betty Sumbler, Hugh Perrett, Richard and Betty Knight, Frank Houlding, Barry Malcolm, Tony Yardley, John Dixon, Allan and Mark Stanistreet, and many more.

As things progressed, we moved on from track-clearance and I first met Harry Lee, who in many ways was the lead inspiration to getting the line running again. By this time John Wood had taken on the role of WSRA Sales Officer, and with Mike Grimoldby set up a sales stand at Minehead to bring in some revenue and to help keep people aware of the preservation scheme.

One weekend we were asked by Harry Lee to turn up at Taunton Concrete Works to remove the jib from a five-ton steam crane. The chassis of this crane is now used as a boiler carrier at Williton. Not long after it became known that the railway, through the WSRA Stock Fund, was interested in purchasing locomotives from the Woodhams Barry scrapyard. Nos. 7820, 4110 and 9681 were the original choices but ended up being Nos. 4561, 5521 and 5542, members of classes of loco that had been regulars on the Minehead branch. In many ways, No. 4561 has been my favourite WSR loco, and I looked forward to its return to steam. On one road trip to Barry I remember tea being passed between vehicles on the motorway, then on arrival fish and chips, a couple of pints of beer and a night's sleep in the WSRA caravan. Officially, the work in Barry was to prepare our locos for removal but in practice this seemed to involve removing parts from other locos in the yard that would be useful. By this time, anything non-ferrous was long gone and earlier groups had removed things like springs and inside motion from our locos.

By August 1975 the three locos were at Bishops Lydeard. The railway then purchased a turntable and associated fittings from Pwllheli, a work week was planned, and off we went with Ray Lee and the WSRA lorry. Accommodation hadn't been planned and we slept in the old goods office near the station. We located the turntable but very quickly realised that our meagre tools weren't going to move it, so we concentrated on water tanks and columns. One standard GWR water tank in the old loco yard now stands in Minehead loco yard. On the Sunday, with no train service running, we drove the lorry along Pwllheli station and set about removing the tank which involved burning off the retaining bolts at ground level then pulling the column over on to a bed of sleepers. The tank was then rolled along the platform like a giant cotton reel past the entrance barrier to be loaded on the lorry. We drove away feeling pleased with our activities only to be awoken the following morning by the station staff telling us that the signalman's water supply had been cut by our activities and could we put it back on! This tank was installed at the Dunster end of Minehead platform, but it now resides at Welshpool, a lot closer to home.

In March 1975 at Watchet I witnessed the Class 25 No. 25229 rumble through – the first train movement for more than four years. We discovered this was to do with the removal of No. 6229 *Duchess of Hamilton* from Butlins camp at Minehead to Swindon works and an eventual return to steam. About a week later, No. 25079 and a couple of brake vans arrived at Minehead. No. 6229 was loaded on to a lorry at the holiday camp and slowly brought to the station, heavy metal plates being laid to spread the weight. To allow unloading, part of the run-round loop was removed in the Main Platform side, something that was to cause a few problems in the next year or so. Some of us followed the train to Blue Anchor then Washford Bank, where we stood at Old Cleeve bridge and waited to see it pass under. The first attempt at slow speed stalled before Bilbrook, the second got a little further but again stalled, the third attempt was at what appeared to be full throttle and got them to Washford. This was my first but by no means last experience with adhesion problems on Washford Bank!

Another memorable day out was to Milverton where the station was to be bulldozed to make room for a by-pass. Shed doors, roof slates and GWR fencing all found their way back to BL. With a Light Railway Order in place and the lease of the line transferred to the WSR, the Bagnall loco *Victor* at Taunton was prepared and in late December it arrived on the branch with the Fruit C van we had been painting two years earlier, plus a hired BR brake van. *Victor* was lit up for the first time and the first WSR train was run to Minehead, carrying company directors, WSRA committee members and invited guests. All of West Somerset appeared to have turned out to see the train through. I then had the privilege of travelling back to Bishops Lydeard on the return working, and after a pint in the local, a bus ride back to Minehead.

David Butcher, newly-appointed as General Manager, and Nick Jones, Operating Superintendent, with Harold Blackmore on the Permanent Way, realised we had an enthusiastic group but limited in railway knowledge, so Mutual Improvement Classes were organised at John Wood's family business in Wellington, where the rudiments of signalling and other operations were explained. Then at the start of 1976 came the long-awaited hand-over of the keys to Minehead Station. Led by retired taxi driver Wilf Sly and his trusty claw hammer, a few of us with former railwayman Don Spencer

In 1975 John Wood and Mike Grimoldby are seen with the railway sales stand at the entrance to Minehead goods yard. Mike's blue J4 van itself is now a vintage vehicle. Allan Stanistreet

began removing boards from the doors and windows to see what had been left. Wilf was to become a stalwart over the next few years renovating buildings and painting coaches.

The state of the station buildings varied considerably: Washford and Crowcombe had been vandalised while others were hardly touched, particularly Williton thanks to being overseen by Edward Martin and his family. With the opening of the first section of line between Minehead and Blue Anchor, rudimentary facilities were arranged at Minehead, with the former carriage-filling hose points brought into use for watering locos and carriage toilet tanks. The former parcels office, under the stewardship of Linda Butcher, was transformed into a shop and buffet and it wasn't long before her sausage stew was on the go to feed a hungry workforce!

Blue Anchor signal box was fairly intact, although a large amount of work was required to get the mechanical equipment working, some it having been disconnected following the movement of *Duchess of Hamilton*. A new ground signal was needed for run-round moves, while S&T work was required at Dunster Sea Lane. By the closure of the line in 1971, Dunster to Minehead had been operated as two single lines with loops at Minehead, which was fine when Dunster Signal Box was open, but as the WSR had decided not to reopen the box the old down line was clipped and padlocked out of use. That meant that what we now know as the Bay was the only siding. To get vehicles from one platform to the other involved a trip to Dunster and back. And until the restoration of the run-round loop at Minehead, shunting had to be done either by hand or as a last resort with chain or wire rope.

We were helped at the time by a Manpower Services grant which allowed the railway to employ out-of-work local people on track work and undergrowth clearance. At the same time items of rolling stock started to appear, notably the Somerset and Dorset Railway Circle vehicles from Radstock plus the three camping coaches removed from Blue Anchor in January 1971 which had been in store at Taunton. These were brought to Minehead for use as staff accommodation.

As the March scheduled start of services approached we had one working steam loco *Victor* and just two coaches, but No. 6412 was bought from the Dart Valley Railway and arrived three days before reopening along with an Auto Trailer hired to provide a brake vehicle if needed. Fortunately, BR was selling MK 1 coaches so BSO 9278, SK 24307 SO 4814 and TSO 4346 were purchased. With newly-arrived full-time volunteer Chris van den Arend riding on board, they came on to the branch and were hauled to Minehead in one memorable train along with No.6412, *Victor,* D7017, and two Park Royal DMU sets. By the skin of our teeth we reached the day the first train ran: Harry Lee drove *Victor,* Don Spencer acted as guard, and new recruit Jim Elson was signalman at Blue Anchor. Chris van den Arend and I spent the day selling platform tickets, but we did get a trip to Blue Anchor where, with the restricted signalling, it was necessary for all passengers to alight on the up platform then go to the down platform while the loco ran round, propelled stock towards Washford, then returned back to the down.

So we had done it after five years of effort and despite hopes being raised then dashed as to when the reopening actually would take place. Little did we realise we had done the easy bit! We were now committed to operating to a timetable and faced some very difficult times. We are still learning as we reach our 44th season!

Salad Day Stints

Gil Bird

My first contact with the West Somerset Railway was in 1973 when I was 16. It provided something interesting to do that was independent of my parents during our annual family holiday in Somerset. Over the next few years I became involved as far as I could. When I was a student, I would spend the whole of my long summer holiday working on the railway. I was accommodated with other volunteers in the camping coach at Minehead. Facilities were best described as basic: there was one "dormitory" complete with bunks, and two highly-prized single berths with the privacy of a sliding door. We could, and did, cook for ourselves, but nothing fancy. The shower had a floor constructed out of a substance not unlike papier-mâché. For other personal plumbing-related needs, we had to use the station amenities.

Before the line was operational, I was involved in weeding the track at Crowcombe and at Minehead, to the strains of Paul McCartney proclaiming *Band on the Run* from the sound system of the neighbouring holiday camp. I was also used for various painting jobs – stations, signals, rolling stock and more. I was mortified to miss the official opening of the railway as far as Blue Anchor in 1976 because it clashed with a compulsory field trip to the Isle of Man organised by my polytechnic. However, during the summer, I fulfilled various roles. The railway was not as professional an outfit as it is now, and its volunteers were encouraged to turn their hands to most things. Apart from the everlasting painting and weeding, I helped out occasionally in the booking office and signal boxes, worked as a Travelling Ticket Inspector and was trained as a fireman and a guard.

Motive power in the railway's salad days was the Pannier Tank No. 6412 and the two industrial saddle tanks, *Victor* and *Vulcan*. We discovered that, in the intense heat of the firebox, a jacket potato could be cooked to perfection in a mere six minutes – and it tasted better than from a microwave! Occasionally I had to light up *Victor* early in the morning to allow the driver to have a more leisurely start. On one occasion I was a bit too keen as well as inexperienced, and the engine was ready to steam an hour earlier than it should have been.

The summer of 1976 is remembered for being long, hot and very dry. The line near Blue Anchor was particularly vulnerable to catching fire with stray sparks from the locomotives. It became part of our routine to stop trains and extinguish any small fires that had broken out on the embankments with beaters.

The summer of 1977 was memorable because of the arrival of a film crew and cast of actors from Southern Television, one of the large independent companies that made up ITV. They were filming the second series of *The Flockton Flyer*, a children's drama about a preserved railway. Those of us working on the railway at the time, as well as operating the "star" locomotive, No. 6412, were also employed as extras. My one and only TV performance to date was as a British soldier at the time of the Indian Mutiny. I sported an adhesive moustache, red tunic and hat with white neck covering. As well as good money, this was enormous fun and fostered a great sense of camaraderie. We appreciated the bacon butties that were served, although the personal attentiveness to intimate details of our costume was less welcome and a source of ribaldry among us.

On one occasion I was firing No. 6412 for some shots near Crowcombe. The TV company wanted an aerial view of the locomotive and hired a cherry-picker to provide the necessary height. No doubt this seemed a novelty to its employees, who crowded into the equipment to admire the scene. As we drove the locomotive into shot, we were astounded to see that the hired vehicle was slipping backwards in the mud. As it slid elegantly from view, the whole thing became unbalanced and tipped over. Its occupants were fortunate indeed to escape with nothing worse than a single fractured arm between them.

Gradually the line was opened as far as Williton. The first trains to run there had a long stop at the station during which time I got on with some painting on the station. There were also forays into the fields towards Stogumber to harvest mushrooms. I guess that I was viewed as a useful pair of hands by the company and, by this time, I was paid expenses for my work over the summer holidays. Most of this found its way into the coffers of local hostelries...

When acting as guard, I needed to report any late departures from stations together with the reason. Once, a small dog failed to "mind the gap" and fell between the platform and train at Watchet. Naturally there was a delay while the animal was retrieved, although I can't now remember how I actually explained this late running.

Some services consisted of diesel multiple units which required less staffing to operate as well as

The Park Royal DMU, W50413 and W56169, was an early mainstay of the line and used to support the steam services. It was acquired with the intention of running the commuter services to Taunton which sadly never happened. The unit was then rebranded as a "scenic railcar". During 1978 the railway extended services to Stogumber. Here the unit is seen preparing for the return trip to Minehead. Allan Stanistreet

being quicker to get underway. Most of the passengers were visitors to Somerset and unfamiliar with the local village names, as well as the trains themselves. I remember being asked, when I was in the booking office, about the actual altitude attained by the diesel train, slightly euphemistically billed as a "scenic railcar", because, the passenger explained: "My wife doesn't like heights." Trying to keep a straight face, I assured the gentleman that the terrain was not actually mountainous. I also sold tickets to people asking about Crewkerne and Stromberg stations...

As well as summer migrants such as myself, railway volunteers were drawn from local people who might do a regular stint. On Wednesdays one retired gentleman acted as Travelling Ticket Inspector. In those days the railway didn't avail itself of customer service training. I used to cringe in the guard's van as I could hear an accusing bellow from the end of the carriage: "Somebody here hasn't got a ticket!" The booming voice had even been known to intimidate more senior members of the railway staff. This same gentleman was reluctant to complete his end-of-day return for fear of making a mistake in reconciling the cash taken against the tickets sold. Like so many on the railway, he was a lovely guy who only had its best interests at heart.

In 1979, the railway opened fully to Bishops Lydeard and early trains weren't without incident. It was a fairly regular occurrence that the locomotives would run short of water. On such occasions the only thing to do was to stop and let the thing cool down. Once the locomotive was rescued by a Somerset Fire Brigade water tender. This was not popular with the fire crew because they didn't see that coming to the aid of an inexperienced and incompetent preserved railway was part of their role.

Trains that summer were only scheduled to travel as far as Stogumber. As guard I would enquire from the passengers if anyone was interested in travelling on to Crowcombe, at no extra charge. Mostly they were, because what they wanted was a steam train ride rather than a particular destination. Once we got to Crowcombe, we had a wait before embarking on the return trip. The whole thing took on the atmosphere of an outing and picnic.

On Saturdays passengers could ride as far as Bishops Lydeard and complete their journey to Taunton on a bus hired to provide transport into the county town. The service buses were operated by, for an obscure historical reason, members of the National Union of Railwaymen, which took a dim view of preserved railways and were most unco-operative towards the special bus. So it was hard for those arriving at Taunton Station to find exactly where it ran from, and at what time. Such was the power of trades unions in the 1970s that they also made ambitions to run a service by rail into Taunton Station via Norton Fitzwarren impractical.

The summer of 1979 was my last on the railway. In that September I started my “grown-up” job as a teacher in Somerset. Despite the long summer holidays there was less time available than when I was a student. I found the workload of teaching hard and, by this time, I had a long-term relationship, for some time over a distance, to maintain. Then I became involved in the restoration of the GWR Prairie Tank No. 5542, but that, as they say, is another story...

Sometimes the two Bagnalls were both running at the same time...here the sisters (brothers?) Victor *and* Vulcan *are seen hauling the siphon van and the Pullman cars into Bishops Lydeard, Easter 1980. David Butcher and Arthur Young are the drivers.* Steve Edge

Flockton Flyer: Partial Saviour of the WSR

Nick Jones

I came out of my office, the booking office shared between its original purpose and me as Operating Superintendent, on to the platform at Minehead one day in late August 1976. David Butcher, the General Manager, was with two people. He introduced me to Colin Nutley and Christine Richardson, who had come from Southern Television to scout locations for a forthcoming children's TV series about a family that resurrected an old steam locomotive and railway branch line. With miles of disused track and No. 6412 with which they'd fallen in love, the WSR was a great choice for them.

We later heard confirmation that Southern TV wanted to use the WSR for a three-week location shoot and that they would pay good money. We could offer almost free-rein on the stretch of railway beyond Williton, and plenty of time between trains for filming at Minehead and elsewhere on the working section of line, particularly at Minehead. A few weeks later a film unit appeared, complete with the cast and, most important, a location canteen. Filming took place towards the end of the longest, hottest summer that most of us had known. They wanted the weather to continue but it broke during the weeks they were with us, resulting in several delays "waiting for the light" or because of rain interruptions.

The star of the show was WSRA locomotive No. 6412, named for the programme *Flockton Flyer*. The Carter family was involved in the restoration of the loco and getting the Flockton to Lane End railway back into use. Filming for one episode at Minehead, which had become Flockton, was attempted on a Saturday. The story was about a train robbery, I think. The Special Projects Team of volunteers led by Hugh Perrett were working with a JCB in the loco shed, digging out spoil so that an inspection pit could be built. The film crew were a bit upset that the noisy work couldn't be stopped. That was just the way it was – the WSR volunteers were giving up their time to push the railway forward.

Scenes for another episode were filmed at Minehead at night. The station was brightly lit and looking really good. In one scene, Bob Carter (played by David Neal) – *Flockton Flyer's* driver – was to walk down the platform brandishing a torch. In his soft-soled sneakers, the scene just didn't sound right. So WSR General Manager David Butcher came to the rescue with the loan of a pair of more suitably noisy shoes.

Some of the filming for the final episode was at Crowcombe. A crash and derailment of No. 6412 was staged on the disconnected down loop line. Any slip by Harry Lee at the controls of No. 6412 could have seen her really dropping off the road, but all turned out well and the TV people were happy with the result. In the same story, the script called for the re-railing to be done by an elephant borrowed from a travelling circus. When the animal arrived for filming, it seemed far too small and looked ridiculous. So, after a hasty re-write a team of working horses was used instead.

Much of the filming took place on the Williton to Bishops Lydeard section, which was still out of use at the time. Looking now at the films, which are available on DVD, gives the viewer a chance to see the condition of the line at the time, looking quite dilapidated. The surrounding countryside, baked though it was after the long drought, looks spectacular as you would expect.

The money provided by the filming contract was an essential part of keeping the railway going. I recall that the original plan was that the facility fee would be £6,000. But as the filming took a fair while longer than anticipated, the final amount was a bit higher. But as the film crew finished, the management and financial difficulties of the WSR were surfacing. A very difficult time followed although I had some personal cheer to be invited, along with a few others involved in the filming, to the Southern TV studios at Southampton to see an episode being put together from the filmed location material and studio videotape. It was fascinating to watch the process and spend time

GWR Pannier Tank No. 6412 poses as the Flockton Flyer while shunting at Washford, 22nd May 1977.
Peter Triggs

Filming the Flockton Flyer. *Harry Lee (far right) looks on as Stan Phipps (sound) and Joe Hardy (camera) prepare for action on No. 6412.* Keith Hawkins

again with the people involved. As it was just before Christmas 1976 we were afterwards invited to a dinner at a nearby restaurant with the producer and other members of the team and cast.

When the series was broadcast by ITV, starting in April 1977, several of us gathered in the buffet at Minehead to watch the final product, and see our impressive railway and countryside being made famous on the small screen. The publicity-conscious new directors of the WSR used *the Flockton Flyer* in a well-organised campaign to bring in the public for the season, and many new admirers for the railway were gained.

Then came news that Southern TV wanted to make a second series. Of course No. 6412 would be the eponymous star again. One of the Bagnalls was also wanted, which presented a bit of a problem. *Victor* was out of traffic long-term, and *Vulcan* hadn't been worked on since arrival at Taunton from BMC Longbridge four or five years earlier. Only a few weeks were available until the start of filming and there continued to be no real facilities for loco restoration. But somehow, largely working in the open, Harry Lee and his team managed to do it. *Vulcan* was put into working order in a matter of weeks and tested. It had to be turned for filming purposes, which involved getting a low-loader into the yard at Minehead, loading the loco in steam, out on to The Esplanade, turning, and reversing back into the yard then rolling the loco off.

Vulcan was to participate in a race with *Flockton Flyer*, the outcome of which would decide the fate of the fictional railway and whether it would fall into the hands of the dastardly Duke of Flockton, played by Patrick Mower of Special Branch and Target fame. The race scenes were shot over a few days on the double track section between Bishops Lydeard and Norton Fitzwarren.

Another episode and several of the storylines in the second series reflected the real-life WSR. It featured a film crew visiting the Flockton line to make a movie, ostensibly

The WSR volunteer platoon assist Christopher Bell (John Moulder-Brown) Bill Jelly (Geoffrey Russell) and Commander Frost (Anthony Sharp) in an episode of the Flockton Flyer where the unlikely plotline revolves around the visit of a film crew to make a historical re-enactment of the Indian Mutiny.
Nick Jones

about an event during the Indian Mutiny of around 1857. Who cared about historical accuracy? There were goodies and baddies, and a fair damsel to protect. Several of the eligible-age staff of the WSR were invited to be in the episode as British soldiers. This was great fun for a few days, even if the First World War-era Vickers machine gun was out of place and very noisy.

On the first day of filming, a number of the crew were in a cherry-picker about 25 feet above rail level at Roebuck Gate Crossing. The truck on which the hoist was mounted toppled over, throwing occupants into the field below, causing some injuries which needed hospital attention, and further filming was delayed. When the crew returned an episode they'd planned involving an air-sea rescue helicopter and Minehead lifeboat had to be cancelled and replaced with a hastily written, simpler story.

When we saw the Carter family and friends steam off into the distance under Heathfield Bridge at the end of the final episode we were sure, as were the crew, that they'd be back next autumn to make a third series, but it wasn't to be. Southern TV's priorities changed: they had a hit on their hands with Worzel Gummidge and that is where resources were concentrated.

Both series were repeated on terrestrial TV, followed by a few outings on cable and satellite. The master tapes were archived and despite several changes of ownership, and some difficulty in tracking them down the programmes were released on DVD so that they can still be enjoyed, showing what the WSR looked like in those early years.

I believe the WSR was, partly, saved by the children's TV series. Without the income from the filming facility fee, there might not have been enough money in the kitty to keep the railway going in 1976 – 1977. Maybe, on viewing in modern times, the stories and programmes are a bit dated and twee, but for the part they played in those early years, I really appreciate the series, and I'm glad that Southern TV chose the WSR.

Re-opening to Bishops Lydeard and Norton Fitzwarren

Nick Jones

While the expectation in 1976 was that the reopening to Taunton would happen in fairly rapid stages, events took a different turn. The reopening of the six miles to Williton from Blue Anchor were achieved quickly enough, ready for the August Bank Holiday weekend after the hottest, most sunny summer that anyone could remember. Then came the first WSR money and management crisis. Accusations were aimed against we professional railway staff by the directors that we were spending money irresponsibly at a time when cash was very short of the projections to which we had not been party. The board announced that the railway would not continue to run the public transport service through the winter of 1976/77, and that all paid staff were under threat of redundancy.

It was the directors who found themselves on the wrong side of this argument and they all stood down the day after an acrimonious AGM, and new directors came in their place. Several redundancies followed, unfortunately. The number leaving was reduced because some of us were able to take a pay cut and others agreed to work for a while without pay. A limited service was resumed for Christmas 1976.

The cash shortage continued, limiting the WSR's ability to expand towards Taunton. Standing in the way were the considerable obstacles of required works: conversion of the two level crossings at Leigh Wood and Roebuck Gate, which had no electricity supply; alterations to the layout and provision of a ground frame and more at Bishops Lydeard; track and drainage improvements in many locations; restoration of cutting and embankment profiles where movement and slippage had occurred; removal of vegetation; fencing. The list went on and on. More money was needed than the WSR had at its disposal, while there was the question of activating a Running Powers Agreement so that WSR trains could run on BR track between the boundary at Norton Fitzwarren and Taunton. The question of motive power and rolling stock availability hadn't even been addressed.

Opening day to Bishops Lydeard, 9 June 1979, sees Jim Summers on platform 2 watching as the Park Royal arrives with Vulcan *sitting in platform 1.* Steve Edge

Into the mix of difficulties was added one more: the National Union of Railwaymen who for historic reasons represented bus staff in the area as well as railway staff, decided that the resumption of a train service between Minehead and Taunton would be a threat to their members' jobs. Accordingly, the WSR was "blacked" and their members would not co-operate with any train movement to or from the WSR. Momentum of the full re-opening project was lost. Somerset County Council, who had purchased the line and surrounding land from BR and let the minimum needed for operation to the WSR, began to take an interest in the still disused part of the line south of Williton for road improvements. There is no doubt in my mind that only half of the present WSR would have survived unless we showed we remained serious about using the whole line.

As a demonstration of intent, it was decided early in 1978 that the trains should start running again between Williton and Stogumber, a section which, as it happened, did not need much attention. The biggest single job was the removal of the disintegrating timber section of platform at Stogumber. In time for the 1978 summer timetable three of the Minehead-Williton DMU services were extended on three days a week to run to and from Stogumber. There was also a proposal from the WSRA to run steam trains at weekends between Bishops Lydeard and Crowcombe. The practicalities of this were quickly identified as unattainable: that we didn't have enough rolling stock or motive power were two of the more obvious reasons, but that didn't stop the printing of a huge number of day-return tickets between the two stations.

By late 1978 the finances of the WSR had stabilised sufficiently for the expenditure on completion of the re-opening of the branch to be authorised. The budget for the project looks rather pathetic in hindsight, but it was all that was available and the directors launched another share offer, which helped the cash situation a little. Those of us on the payroll had become used to not receiving our pay on time or at all, anyway. The budget for the conversion of the level crossings was £10,000 (around £55,000 in 2019 value), including the provision of a power supply. Our solution was to have mains connected to Crowcombe station (there was no supply during BR days) and to run an underground cable to both crossings, a length of just over a mile-and-a-quarter, pretty much on the limit of what could be achieved with the materials and technology then available to us.

Roebuck Gate and Leigh Wood crossings were only a small part of what had to be got ready, so I was tasked with applying for a Government grant to help us along. We asked for up to 60 short-term employees funded by taxpayers to assist with the same sort of jobs that had been undertaken by the Job-Creation programme back in 1976. We gained approval for a grant, including an amount for tools and materials. Jobs like the repainting of the bridges at Watersmeet, Tribble, Combe Florey and Sampford Brett were in the programme as well as remodelling the track at Bishops Lydeard and making the station buildings at Stogumber and Crowcombe wind and watertight. These tasks were in addition to much removal of vegetation, track fettling and repairs to fences and locations where earth slips had occurred. A start was made early in February 1979 with the target of getting the work done for a grand re-opening on 9th June 1979. At the same time we had a train service to run. Getting the whole line between Stogumber and the boundary at Norton Fitzwarren up to scratch, turning back the results of decades of BR neglect, was a great strain on our slender resources.

There was a massive contribution from volunteers of the WSRA. Harold Blackmore and John Sumbler were the entire company staff of the Permanent Way Department and without the volunteer gang led by Graham Harwood the project would have taken much longer, or failed. The same can be said for the volunteers responsible for the installation of the level crossing signalling, principally John Jenkins and Jon Francis who brought their experience on BR with them. Without doubt, the task was massive and the number of people involved was very small. The board had asked me to lead the project, which became a seven-days-a-week obsession for a few months.

Harry Lee, in the Locomotive Department, was busy trying to ensure there would be three working steam locomotives ready for the reopening. The loco fleet consisted of the WSRA's 0-6-0PT No.6412 and the two Bagnall 0-6-0 ST engines, *Victor* and *Vulcan*, purchased by the WSR in

1976. *Victor* was readied for the 1976 re-opening to Blue Anchor and *Vulcan* was back in traffic for the autumn of 1977, but the two were rarely in traffic at the same time since one or other, or both, were out of traffic for repairs frequently. They were heavy industrial locomotives designed for low-speed heavy work over short distances, not really suitable for the WSR's distances or the required speeds.

On 31st May 1979 Major Peter Olver, of the Railways Inspectorate, came to Somerset for a two-day inspection to approve (or not) the work done on the level crossings and the entire line for passenger trains. Following criticism from an adjoining landowner about a perceived lack of rigour at an earlier inspection, we got the full treatment. The second day involved a track walk all the way from Stogumber to Norton Fitzwarren. Major Olver declined the offer of a DMU ride, stopping at locations of interest for a closer look, and instead insisted on the long walk accompanied by Harold Blackmore and John Sumbler (PW), John Jenkins (S&T) and me. He had a train to catch at Taunton to take him back to London so we started early, striding out over the sleepers and stopping where he wanted to take a closer look at a culvert here, a bridge there. Around lunchtime we were near Combe Florey where Harold Blackmore suggested we should go down the embankment to the Farmers Arms. It was a fairly hot day, and it was a reluctant return to the track walk after a liquid lunch and, for me, a banging headache. There was no stopping the ex-Royal Engineers officer who strode on making notes using his portable recorder.

It was a great relief to arrive at Norton Fitzwarren, and to hear him declare that we had his approval, clearing the way for the planned reopening on 9th June. There were many odd jobs to tidy up but we had achieved what we had set out to do. I was very pleased and relieved, and I slept well that night. Back at Minehead things hadn't been going quite so well. Ideally we needed three steam locos to run the planned gala re-opening but only one, *Vulcan*, was in a fit state, No. 6412 had long-term problems, and there was a struggle to get *Victor* going in time. Early in May, I had been with Harry Lee to Peterborough to see a couple of ex-BR Mk1 non-corridor suburban coaches from the Nene Valley Railway which weren't ideal but there wasn't much option as we urgently needed more than six passenger coaches. On seeing they were basically intact, we agreed the purchase.

The re-opening gala was ambitious with six trains each way on the Saturday and seven on the Sunday, plus an evening picnic special on both days. The DMUs, steam locos and D7017 all needed to be in action. To say things didn't go to plan would be an understatement – despite round-the-clock work, the steam loco situation had not improved. Early in the morning I phoned HM Railway Inspector asking if we could use the Hymek more than we had planned. It was over the line's axle load limit of 17 tons 12 cwt and could only be run with HMRI permission. Thankfully it was agreed but even then the re-planned service didn't run well. We had overreached ourselves. It wasn't a great start to the extended railway but we had achieved the re-opening.

My abiding memory of the Saturday is of D7017 failing with the return picnic special, which had *Vulcan* simmering in the consist not intended to work. In a super example of enginemanship, the steam loco crew of Harry Lee and Ray Lee got *Vulcan* back up to working pressure to start on the steep bank, pushing the Hymek and bringing the tired passengers back to Minehead. Quite an achievement of which all concerned, volunteer and paid, full-time and temporary, were rightly very proud.

Early History of the West Somerset Railway

Allan Stanistreet

The Minehead branch of British Railways (Western Region) was closed from 4th January 1971, being the last of the branch lines radiating from Taunton to do so. However, even before closure, plans were afoot to reopen the line under private auspices, with the aim of running a diesel-hauled commuter service to Taunton, supplemented during the holiday season and some other times by steam-hauled trains. For various reasons these grand plans never came to fruition, but after five years of negotiations, the line did reopen with more limited ambitions. Unusually, British Railways were persuaded not to remove the rails immediately, pending plans for reopening, and thus we see the highly successful tourist line of today.

It is now almost half-a-century since the line was closed by BR. Just four months later, the West Somerset Railway Company was formed and at roughly the same time the supporting West Somerset Railway Association (WSRA), not then a charity, was inaugurated. The aim of the association was to support the company in its endeavours to reopen the line and thereafter to provide volunteers to assist in the running of the railway. Its support was to prove crucial in the years ahead, particularly from a financial point of view.

It goes without saying that many of those who were around in those early days are sadly no longer with us. Those who actually worked on the line and still do are fewer still, with myself being numbered among them. I joined the WSRA in July, 1974, with membership number 208. If memory serves, the chairman then was the late John Pearce, secretary Brian Crudge, treasurer Mike Grimoldby and membership secretary the late George Hitchcock, who had actually worked for the post-grouping GWR. The membership was so small that most of us knew each other.

At that time the nascent company had no staff and most of the work was done by the Chairman, the late Douglas Fear, a Taunton undertaker and photographer, and the Company Secretary, George Byam-Shaw. However, a few volunteers did give occasional assistance. The Association volunteers therefore got on with raising funds by whatever means they could. This involved such things as collecting old newspapers, which then commanded a good price, a sales stand selling books and small railway artefacts, and collecting trading stamps (younger readers may ask: what are they?) to buy such useful items as spades and other gardening tools for practical work along the line, which at that time had been almost completely enveloped in places by jungle.

However, it was to be a few years before we were allowed access to the line for practical work. In the meantime the company managed to acquire two steam locomotives which were stored in Taunton west yard along with a number of locomotives and items of rolling stock, mostly the property of the Great Western Society, Taunton branch. The two two-car Park Royal diesel multiple units were also stored there. Several small industrial locomotives destined for the line were stored at Norton Fitzwarren in the long-departed Blinkhorn sidings. The WSR locomotives were *Victor* and *Vulcan*, 0-6-0 saddle tanks built for a steel works in South Wales but eventually ending up with the Austin car works in Birmingham from where the WSR acquired them.

Access was allowed to the yard so the late Harry Lee, the Association's Locomotive Officer, and his merry men set about putting *Victor* in running order for use on the line. This involved fitting it with a vacuum brake for passenger trains, something it had not required as an industrial shunting engine. Full access to the line was only granted late in the day, about a year before reopening; when it happened, a permanent way department was formed with Graham Harwood at its head to try to remedy many years of dereliction. I then became a member of his team. Like so many lines, the West Somerset had been allowed to run down from the mid-1960s and only minimal maintenance had been done. Although the line was still physically intact between Minehead and Taunton, weeds and undergrowth had taken over. All the surviving buildings had been boarded up, and the railway overall presented a sorry sight.

Bishops Lydeard signal box was an empty shell and the passing loop at Crowcombe had been severed several years before. Williton up loop had been shortened in the 1960s to avoid the expense of rebuilding the culvert bridge as a double line, leaving only half the platform in use. On closure, only three-car DMUs operated the line, not requiring the full platform. Minehead had two tracks, one into platform 1 and one into platform 2, both severely truncated.

The permanent way department set about the task of weeding every station as well as replacing lost keys – which hold the rail to the chair on the sleepers – and oiling all the fishplates. The magnitude of these tasks may well be imagined in terms of a 20-mile line, but it was eventually achieved. Another major task undertaken without mechanical aids was the removal of the siding point at Blue Anchor and its transport to Minehead for use there.

Three locomotives had been purchased from Barry scrapyard in South Wales for eventual use on the line. They were former GWR Prairie tanks, numbers 4561, 5521 and 5542. All three were stored at Bishops Lydeard pending decisions about their use. Small teams led by the late Harry Lee and John Pearce made many trips to Barry to ensure they were properly protected from the depredations of the weather and the attentions of the unscrupulous! The same people also recovered the turntable and water tower from Pwllheli in North Wales, which were surplus to BR requirements.

13 March 1975 saw the first movements along the branch since closure in 1971. Duchess of Hamilton *had been a feature at Butlins since 1964. In 1975 it was moved by rail from Minehead eventually to the National Railway Museum. Here the engine is seen being hauled by class 25 25229 near Roebuck Crossing.* Peter Triggs

The turntable is now in use at Minehead, having been suitably extended. It spent many years being moved around Minehead yard before installation in 2008. The water tower was installed on the platform end at Minehead until that facility was moved to the yard, where it proved unsuitable, being too short, and it now does duty at Welshpool on the Welshpool and Llanfair Light Railway.

One way and another, an enormous amount of work was done by the Association volunteers to ensure that the railway was in a position to reopen on the appointed day. Just before that, the signals and telegraph department under the leadership of Nick Jones gained access to the surviving signal box at Blue Anchor to free all the mechanism, the associated signal wires and point rodding and the level-crossing gates which hadn't worked for more than five years. The large wheel which operated them had been sold off to an enthusiast when the box closed, and he generously donated it back to the railway when the box reopened. Our present signalling engineer, John Jenkins, has been heavily involved with the signalling on the line since reopening, he being one of the few still with us from the earliest days.

The first train from Bishops Lydeard to Minehead under the present company ran on 21st December 1975. It consisted of *Victor*, the WSRA's GWR Fruit C van and a goods brake van. The brake van carried company officials and Tom King, then the MP for Bridgwater and a great supporter of the line. Many hundreds of people turned out on this winter's day to observe the passing of the train and welcome it into Minehead. However, there was still much to be done before the line could reopen to the public.

The former goods shed had been designated as the future engine shed and a line through it was reinstated under the direction of the late Harold Blackmore, who had taken charge of the civil engineering side of things. Harold, known as "Larky", though not to his face apparently, was a retired ganger from BR who had been based at Taunton.

Immediately before reopening, there were still many things to be done such as movement of rolling stock to Minehead to accommodate the anticipated numbers of passengers. Initially services were to run only to Blue Anchor and return. However, until the opening day, the company had no rolling stock of its own and relied on the Association to provide at least some of the coaches. The Association also acquired a further working locomotive with an attendant auto-coach from the Dart Valley Railway where they had been surplus to requirements. These travelled from Devon under their own steam via the main line. The locomotive was an ex-GWR pannier tank No. 6412. In the event, the auto-coach was never used on the line and became a store on platform 2 for such things as surplus carriage seats.

Everything was now set for the opening day, which was announced as 28th March, 1976. It was the original intention to invite the actress Angela Baddeley, a relative of George Byam-Shaw, the Company Secretary, to perform the opening ceremony. Unfortunately, she died just before so at short notice Lord Montagu of Beaulieu, a noted railway enthusiast, was asked to officiate. Watched by several thousand people, he duly waved off the first train. In those far off days, Health and Safety was not even a twinkle in some panjandrum's eye and not only was there not a high-viz jacket in sight but there were members of the public all over the tracks and, to cap it all, detonators were placed on the rails for *Victor* to activate as it left the platform.

The original timetable was not overly ambitious, with only four trains a day from Minehead: 10.00, 12.00, 14.00 and 16.00. The only fares were 50p return for adults and 30p for children. About 15 minutes were allowed in Blue Anchor to cater for the locomotive running round its train. Operation was by train staff, one engine in steam. *Victor* and No.6412 shared the day's trains. Facilities at Minehead were rudimentary to say the least, with water being supplied via a 2ins hose from a stopcock on the platform, which was inserted into the locomotive tank after the 16.00 train had returned, in the hope that the tank would be filled by the time of the next train at 10.00 the following day! Coaling was equally primitive, being done by hand from a five-plank wagon behind the locomotive in the run-round road.

From tiny acorns do mighty oak trees grow! More than 2,000 people were said to have travelled on the line on the first day. Staff were required to wear a maroon coat sporting a deer's head on the pocket, resulting in the attire being described as "a zoo-keeper's outfit!" It was soon relegated to the unwanted clothing store. Part of what is now the Buffer Stop shop was converted into a buffet selling tea, coffee, soft drinks, ice cream and light refreshments. However, at busy times both the accommodation and the staff could be overwhelmed and it didn't last very long, with the entire area being given over to a gift shop, as now. So the newly-opened West Somerset Railway embarked on its first season on a wave of optimism, but there were storm clouds on the horizon.

The euphoria of March 1976 wasn't to last for long and it soon became apparent that running a railway as a tourist attraction wasn't something to be undertaken lightly, even if the distance was only three-and-a-quarter miles. There was an early indication that those responsible had little idea of what it was all about when the tickets ordered for the reopening were not the standard Edmondson size and so did not fit the ticket racks, although they could be date-stamped on the standard dating machines. It was to be some time before they could be replaced by tickets of the correct dimensions.

From day one locomotive availability was a problem, there being only the original two locomotives to run the service: *Victor* and No. 6412. By May 1976, No.6412 was suffering from piston problems, which left *Victor* to do the lion's share of the work although one of the Park Royal DMUs fortunately was also available.

No sooner was Blue Anchor reached than efforts were directed to reopening the section to Williton and considerable work was being done on the permanent way. The view has been expressed in some quarters that, with hindsight, it might have been better to consolidate what was already up and running rather than attempt to progress too far, too fast. However, events moved on and the section from Blue Anchor to Williton reopened on 28th August that year.

The Minehead-Williton timetable initially contained seven trains each way but only three were steam hauled, the remainder being the Park Royal DMU. Departures from Minehead were 09.00, 10.00 (Steam), 11.15, 12.30 (Steam), 13.45, 15.00 (Steam) and 16.15, with departures from Williton at 09.55, 11.05 (Steam), 12.25, 13.35 (Steam), 14.55, 16.05 (Steam) and 17.25. Return fares were: Williton adult, £1.30, child under 14, 60p; and Watchet adult, £1.10, child, 50p.

When trains first returned to Williton, some services were run by the Park Royal DMU. Here the unit is seen terminating at Williton. Tony Yardley

While operating over the relatively flat terrain between Minehead and Blue Anchor was not unduly taxing for the limited motive power available, the climb of Washford bank proved challenging, particularly for *Victor*, originally designed along with its two sisters for shunting extremely heavy loads over short, level distances in a South Wales steel works. Pulling six laden passenger coaches up the 1 in 65 gradient didn't suit the locomotive and it was shortly taken out of service to have three new tubes fitted and was reduced to pulling four-coach trains.

By September 1976, to use modern parlance, the wheel had almost come off. It was announced that the projected winter service would be abandoned and the staff would be laid off until spring 1977. This was unwelcome news for the paid staff, some of whom had left better-paid work for the railway. Although the railway had not performed too badly during the summer, financial problems and difficulties with the Somerset County Council loan were given as the reasons for this decision.

There was some good news. It was announced that No. 6412 would be featured in a Southern Television programme *The Flockton Flyer* which was about a family attempting to rescue an abandoned branch line and reopen it as a tourist attraction. All the outdoor scenes were filmed on the WSR. It was anticipated that this programme would be transmitted during February/March of the following year (1977) and no doubt it provided much-needed funds for the company's coffers.

As a result of the announcement of the proposed winter closure and laying-off of staff, the West Somerset Railway Independent Review Committee was set up. The members were Richard Stevens, a chartered surveyor; Christopher Dowrick, a bank manager; Gerard Belton, a local government officer; and Michael Grimoldby, a company administration manager. He was also treasurer of the WSRA. Mr Stevens was nominated as Managing Director of the committee, while Mr Dowrick was Financial Director. Mr Belton was Marketing Director and Mr Grimoldby also represented the WSRA. Meanwhile, the staff and volunteers continued to carry on working at Minehead and elsewhere.

The company held its first Annual General Meeting on 30th October 1976 at Minehead Middle School and attracted an extremely full attendance. To call it a tumultuous occasion would not be an overstatement. For some reason Alan Garraway, then General Manager of the Ffestiniog Railway, acted as Chairman having been appointed to the board the previous evening at a fee of £50 a day and was to leave the board at the end of the meeting. Douglas Fear, the Managing Director, having been severely criticised for his running of the railway, resigned from the board feeling unwell. As far as adopting the directors' annual report and accounts was concerned, it was only the large proxy vote which carried the day for the board, as almost the entire audience voted against.

Two days later Mr Fear announced that he didn't intend to resign as he didn't see why he should. The board then unanimously decided to terminate his employment and offer him an undisclosed amount in compensation. He declined to reconsider his resignation and he was dismissed from the board, eventually agreeing to accept an undisclosed sum in compensation. A new board was elected consisting of Jim Bennett, John Williams, Mike Grimoldby, Gerry Belton, Richard Stevens and Christopher Dowrick, with Mr Belton as Chairman. Although the proposed winter services would be

August 1977. The loco department led by Harry Lee, rises to the challenge of getting Vulcan *ready to guest star (along with Patrick Mower) in the second* Flockton Flyer *series. The TV crew were due to arrive to start filming on 6th September. Needless to say, despite the lack of facilities, the date was made.* Nick Jones

suspended, all staff excluding the General Manager would be given 28 days' notice of redundancy but every avenue would be explored to retain them.

In due course it was found possible to reinstate a limited winter service and it was announced in December that a DMU service would operate from 11th December, with three return trips on Tuesdays, Fridays and Saturdays between Minehead and Williton. This would continue until the restoration of full services the following March. It was also found possible to withdraw the redundancy notices to all staff except two, and these were voluntary. Gerry Belton resigned as Chairman on 31st December and Richard Stevens was appointed in his place.

In the midst of all these trials and tribulations, work on the engine shed pit at Minehead was progressing. The foundations having been dug on 18th December 1976, three large lorry-loads of concrete were poured for the pit floor. Work commenced at 07.30 and was complete by 11.00. The reinforcing steel for both walls was fixed in place over the Christmas holiday, the steel shuttering erected and it was hoped to pour the concrete for the walls around mid-January. Overseeing this task was "Hugh" Perrett, very much a hands-on man and sometime Chairman of the WSRA. "Hugh" is still an active volunteer, having served the railway now for more than 40 years – a remarkable and most praiseworthy record.

The New Year saw the resignation of General Manager David Butcher on doctor's advice. Despite being in the post for just over a year, he had been the driving force behind the push to open the railway to Williton. Appointed in his place was Major Tony

The sun glints down at Minehead as Vulcan *is seen double-headed with Hymek D7017, owned by the Diesel and Electric Group.* Tony Yardley

Griffiths, who had been associated with the company since the early days. The board appointed Chris Rundle as Publicity Officer.

It was announced that steam haulage was to operate only between Minehead and Blue Anchor, with DMUs from there to Williton. From February 1977 WSRA members' benefits were to include, rather surprisingly in view of the company's fragile financial situation, no fewer than three days' free travel on all public trains in the timetable. However, this offer was to be reviewed at the end of the season.

It was decided that while all existing firemen would retain their status, any new volunteers in the locomotive department would join as cleaners and advance in the traditional way through the grades. It was considered essential for personnel to receive appropriate training for the post to avoid unnecessary damage to locomotives by careless firing.

On 1st April 1977 the West Somerset Rail Users' Association came into being with membership at 50p open to all permanent residents of Somerset and all WSRA members, whether or not resident. Members were entitled to travel on the WSR at winter rates at all times except Specials privately hired. This scheme, suitably updated, has survived to the present day, although it can no longer be had for 50p!

Having gained experience with the lead at Blue Anchor before the reopening, Graham Harwood and his PW gang successfully removed the lead at Dunster, once again without the use of any plant. The completed inspection pit in the shed was handed over to the locomotive department and it continues to give good service more than 40 years later.

Meanwhile, work was forging ahead with a view to extending the service to Bishops

Lydeard. A considerable amount of hard work had been done to bring the three remaining stations up to standard: Bishops Lydeard had long had its own working group, and Stogumber and Crowcombe were slowly being restored to respectability by Walter Harris and Mike Hodge at Crowcombe and Tom Waller at Stogumber. However, much effort was still required to bring the track to an acceptable standard to run trains. The stretch from Williton to Bishops Lydeard had suffered from the ravages of a jungle takeover much more than the section from Williton to Minehead.

For several years only two operational signal boxes remained on the line – at Blue Anchor and Williton. Minehead was still being operated by ground frames, although the up line had long been taken out of use and lifted for use elsewhere. Train operation was by means of the Staff and Ticket system, which worked well enough most of the time but is inflexible and can involve road staff transfers to maintain the service. The locomotive situation was still extremely parlous, with only No. 6412 in service for most of the time. *Victor* spent much of the time under repair, but there was some good news when it was announced that *Vulcan* was now back in service, pulling its first train on 6th September 1977. This must have proved a blessed relief to the loco department who were struggling to maintain a service. Work had started on restoring No. 4561 but it was to be a while before she returned to service.

January 1978 marked the introduction of the first edition of the *West Somerset Railway Journal*, replacing the monthly Newsletter which had been introduced some six years before. The *Journal* was to be a quarterly publication and would be more in keeping with the railway's image. The then Editor found his military duties made it difficult to produce a monthly publication to be dispatched to the 800-plus membership. The new format would still contain up-to-date news of the line and its various departments and it would be illustrated.

The Bishops Lydeard sub-committee decided, with the agreement of the WSRA committee and the company, that the signal box should be restored externally and a maximum 33-lever frame should be installed. A good deal of the work was undertaken by the Willicombe brothers, Ivan and Arthur. John Jenkins and Jon Francis were happy to lend their professional assistance to the project.

Once again only *Vulcan* and No. 6412 were available for the 1978 season, with *Victor* out of service for repairs. In fact No. 6412's boiler certificate expired in May, but the boiler inspector was prepared to extend it until mid-October to enable the company to run it alternately with *Victor* and *Vulcan* until the end of the season. Peckett 1163 *Whitehead* was doing duty as yard pilot at Minehead, being unsuitable for hauling trains due to its small size and lack of vacuum brakes. The Park Royal DMUs gave sterling service and were averaging nearly 20,000 miles a year.

February 1978 saw severe blizzards, with roads blocked for days. The WSR proved its worth when by special dispensation from the Department of Transport, trains were run between Minehead and Bishops Lydeard carrying passengers, Post Office mail and urgent medical supplies. The main road was cleared from Taunton to Bishops Lydeard so a connection could be made. Since the line was not then open between Williton and Bishops Lydeard fares had to be improvised and *Vulcan*, driven by Harry Lee, was used for the service with one coach attached.

A Gala Day was held on 7th May 1978 to celebrate the return of trains to Stogumber.

Since there was no run-round facility, only the DMUs would be used for the service. The rotten wooden platform had to be demolished before trains could run and it was to be some time before the platform was restored to its original length. More than 1,700 passengers made the journey and there were various attractions along the line. The day was judged a great success, bringing a welcome addition to railway funds.

In May 1977, "Hugh" Perrett was asked by the company to inspect the signal box at Dunster with a view to it being relocated to Minehead. The Railway Inspectorate had insisted on its removal or demolition to provide better visibility at the adjacent level crossing. The move was scheduled for Sunday 20th November 1977, and the April 1978 *Journal* contains a fascinating article by "Hugh" dealing with the logistics of this exercise which was believed to be the first of its kind on a heritage railway. As a portent of things to come, the October 1978 *Journal* recorded under the heading "Signal and Telegraph" that the down home signal at Williton had been fitted experimentally with electric lighting to see if the work and expense of oil lighting could be abolished. All signal lights on the line are now electric. Six electric token instruments had been acquired although funds had to be found to pay for the cable. It was to be some years before the first installation between Minehead and Blue Anchor, and several more before the whole line was so equipped.

Late 1978 saw the S & T department removing the redundant frame from Westbury South signal box for use on the WSR. It was intended that in due course 33 of its 99 original levers would be used in the restored Bishops Lydeard signal box. Meanwhile plans were in hand to replace the level crossing gates at Blue Anchor and Williton, where both original sets were showing signs of advanced decrepitude.

The board had given the go-ahead for the installation of colour lights at both Leigh Wood and Roebuck crossings, between Stogumber and Crowcombe, as they were essential before trains could run on the section. These installations were passed by Major Olver, of the Railway Inspectorate, on 31st May and 1st June, paving the way for reopening between Williton and Bishops Lydeard. Undoubtedly the high point of 1979 was the Gala weekend held to celebrate the reopening of the line to Norton Fitzwarren after many years of hard work by a large number of people. This was the first time for eight-and-a-half years that trains had run over the full length of the line and a number of attractions were advertised for the weekend, including an augmented train service, displays of vintage transport, exhibition of local crafts, model railway exhibits, special concessionary fares for Association members, and evening activities on the Saturday. Nearly 2,000 passengers over the two days journeyed over the line. A mini-rally was held in Bishops Lydeard station car park with some stationary engines and, among other exhibits, Brian Crudge's Allis-Chalmers "B" tractor. Walter Harris arranged an extensive Hornby model railway layout at Crowcombe while the S&DJR Trust held an open day at Washford. The weekend was only marred by some indifferent timekeeping and the fact that the available steam locomotives were simply not up to the job. *Vulcan* and No.6412 were sorely challenged by the new length of the line with its curves and steep gradients. In the end, the day was only saved by the Park Royal DMUs and Hymek No.7017 which just about kept traffic on the move.

The year was a watershed for the company's stability and financial standing, and it was to be another five years before light was beginning to be seen at the end of a very long tunnel.

My Early Days on the WSR

Adrian Read

In the early 1970s many heritage railways started to plan possible operation, and as a schoolboy in Taunton I was quite excited that one was about to open so close to home. I used to walk to school trying to think what I could do to get involved. My parents had bought some WSR shares at the initial share offering, but we weren't at the opening day. In fact, I think it was 1977 before I managed to get them on a bus to Williton to take our first trip on the WSR. It was then that I picked up a leaflet and joined the WSRA. I got in touch to offer my services, explaining that I needed transport to get out to the line. I was put in contact with Graham Harwood, who was then organising the volunteer PW gang and had a van for tools and picking up volunteers on his way through Taunton to Bishops Lydeard.

So at the age of 15 my career on the WSR started. I jumped in the back of the van, sat on a few cushions among the tools – rightly or wrongly, such an operation would not be possible today. My first working party was changing sleepers near Watersmeet bridge, just north of BL, and the Wickham trolley was used to transport manpower, tools and replacement sleepers to the site. Sleepers were taken up from the redundant second track south of BL, although some of the double track was being used for filming Southern TV's *The Flockton Flyer* and only some were being lifted until filming had finished.

On that day, like on many other early working parties, we met Harold Blackmore and John Sumbler from the Company gang. Harold had been on Taunton PW since the days of the GWR and he knew everything there was to know about PW, and despite being past retirement age, he was still able to make the heaviest work seem easy. It certainly taught me about the world of manual work and how much could be achieved by hard graft and a gang of people working together. Apart from the Wickham trolley, it was all manual work with no heavy plant to help. Harold had a strong Somerset accent and a certain way with words. He was the leader of the gang; we did as we were told. It was a hard training course but quite an education, which gave us all a lot of

The PW gang take a breather at the north end of Bishops Lydeard as one of the Bagnalls approaches.
Steve Edge

experience that has proved invaluable in the years since. And at the end of the day he always said a cheery "thank you." A top bloke.

As all the Sunday PW gang came from the Taunton area and had their base at Bishops Lydeard, most of my early involvement was at the southern end of the line. Much work was done to get the track fit for some Christmas specials from Minehead to BL before full re-opening. Then work was concentrated ahead of re-opening to Stogumber in 1978, and finally to BL and Norton Fitzwarren in 1979. As well as the volunteer gang there were numerous weekday gangs of Government-sponsored work schemes that were mainly involved with cutting back trees, doing fencing and other necessary tasks. There were also groups of offenders supervised by the Probation Service, but they weren't always the most motivated of workers.

For the re-opening to BL and Norton, the PW gang had the weekend off, so I volunteered to do something to help at BL although I had no experience of station duties or customer care. I turned up on the Saturday and somehow got put in the booking office. Nick Jones was in charge and he gave me a quick introduction to duties. And that was it. There weren't too many volunteers around so I was in the booking office all day. The timetable collapsed as the locos were struggling, and I had to press on as best as I could. It was certainly a busy day. Then back for more of the same on the Sunday. It was quite an introduction to a new job, but without card payments, vouchers or other promotions, it was relatively straightforward. Just very busy, and I had certainly never handled so much money! And I still wasn't old enough to vote...

In the early days of the WSR there was a real ambition to provide a public service, and on Saturdays a DMU service ran with a connecting contract bus to provide a link to Taunton station and town centre. This provided a service both for people to visit Taunton and for holidaymakers heading for West Somerset. I took it on myself to help out by riding on the buses, advising passengers and helping with luggage. Initially the service was operated by Hardings Coaches who used a variety of vehicles and drivers. In later years, Berry's ran the service using Bristol Lodekka double decker buses that were used for school duties on weekdays. For many years these were mainly driven by Ken Harris, another solid bloke who took pride in his work and who also had a splendid broad Somerset accent. These buses carried a big advert for the WSR painted on the side, which gave good publicity.

Despite these diversions, it was the PW gang that was my main focus for railway work. With the line growing and other activities taking place, several volunteers went from the PW to other things so that sometimes the Sunday volunteer gang was reduced to only a handful of people. On one occasion we arrived at Churchlands bridge near Combe Florey to see smoke coming out of the Farmer's Arms inn. We were there before the fire brigade and helped remove furniture from the bar area to the car park.

Occasionally we had trips off the WSR, such as to the Bristol Harbour Railway where we removed some point-work from the dockside which was later installed at Williton. We travelled to Montacute over several weekends to salvage the remains of the station ahead of the site being buried under the new bypass. Precast concrete sections were taken away and used to build the new Doniford Halt.

During some school holidays I joined the WSR PW gang with Harold, John and others and I remember Bruce, from Carhampton, who was a bit of a laugh. Week days were even harder than the Sunday gang: I was picked up at 7.30am, we stopped for breakfast for 15 minutes at around 10 o'clock and for 45 minutes for lunch at about 1pm. Generally on Fridays the whole line was walked, points oiled, fish-plates tightened as necessary, and any faults identified. Other work was often mainly keeping a good top and line to the track to give a smooth ride, all with hand tools. Unfortunately, the Bagnall saddle tanks then in use gave a rough ride for passengers and were equally rough on the track.

Adrian at work in Bishops Lydeard ticket office on opening weekend 1979. This is now the mess room on platform 1. Steve Edge

So much has changed since. There is a much greater emphasis on safety, more staff, and a lot more heavy tools and equipment. The variety of locomotives is beyond the wildest dreams of the early days, as is the length of trains and number of passengers. I can't begin to imagine what the next 40 years might bring!

Bishops Lydeard had to be transformed from a quiet country station to the southern terminus of the line. Above, it is seen before trains returned in 1979, showing the basic facilities and the amount of work that needed to be done to get it ready for passengers. The United Dairies shunter sits in the goods siding. Allan Stanistreet

Below, before the water tower was available, this shows how the steam locomotives had to be refilled in 1979. Tony Yardley

Saving the Prairies

Tim King and John Wood

John Wood is the Chairman of Locomotive 5542 Ltd and Tim King is a founder shareholder.

When Warner Bros was looking for the perfect steam locomotive to star as The *Hogwarts Express* in the Harry Potter movies, it was a GWR loco that got the part. In every movie it graces Platform Nine-and-three-quarters at King's Cross and chuffs its way along various Scottish lines, heading for the famous Hogwarts School. However, the cinema-goers who between them paid more than $7 billion for their tickets are unlikely to know the real story of this now familiar locomotive and how it, and hundreds more, narrowly escaped a much darker fate.

John Wood on the footplate

Just under 17 miles across the Bristol Channel from Minehead is the port of Barry. This once-busy dock complex and rail yards used to be brimming with trucks of coal waiting to be exported by sea to every corner of the Empire and beyond. After the demand for coal declined, the sidings took on a new role: under the 1955 Modernisation Plan for British Railways, thousands of wagons and 16,000 steam locomotives were due to be taken out of service. When BR's engineering works at places like Swindon and Eastleigh became overwhelmed with the task of cutting these up, tenders were issued to rail-connected scrapyards around the country. Dai Woodham ran a well-established family scrap business in Barry and did a deal with BR that eventually saw 297 locomotives and thousands of wagons and brake vans filling up the sidings in Barry. For 20 years the town became a Mecca for steam enthusiasts as the locomotives slowly rotted in the sea air. Dai kept his men busy cutting up rails and wagons and left the steam locos for a rainy day.

Within the ranks of these engines were GWR No. 5972 *Olton Hall*, the future *Hogwarts Express* and a wonderful selection of GWR examples from Kings to humble Pannier tanks. With new heritage railways opening up around the country, the preservation pioneers needed steam power and they headed to Barry with a view to purchasing these rusty old hulks to give them a new lease of life.

The three prairie tanks are seen here being offloaded at Bishops Lydeard. Peter Triggs

In 1974 things on the West Somerset Railway appeared to be progressing towards reopening, but frustratingly for those involved on the ground, only at a very slow rate with seemingly delay after delay. The only real progress being made was on basic pick and shovel efforts to spruce up the trackwork in the station areas and some limited fund-raising. However, if the railway was to succeed in the long term it would need a supply of suitable steam locomotives to run its tourist services. In those days the plan was to operate the railway in two distinct parts: regular commuter traffic to Taunton, and tourist steam railway. Two operational Bagnall steam locos had been secured from British Leyland at Birmingham and would help get things underway, but more suitable locos were needed if the planned services were to stand a chance of success. Consideration was made of various locomotives remaining at Barry, but the final decision was made that three small Prairie tank locos would be ideal. The class had been seen regularly on the branch in Great Western and British Railways days. Locos of the same class would give the benefit of interchangeability of parts and should the need arise for heavier trains, then two could work together. Much surveying of the locos available had been carried out a few years earlier when the Taunton branch of the Great Western Society had purchased No. 5572.

Members of the West Somerset Railway Association's Stock Fund visited the yard at Barry to undertake final inspections, and eventually did a deal with Dai to buy three Prairie tanks – Nos.4561, 5521 and 5542. Funds were raised from the ranks of WSRA members and the public. Regular working parties took place at weekends for almost a year, with 20 or more volunteers making their way to Barry with tools and equipment to collect parts and make the locos ready for transport to Somerset. In September 1975, the three locos were triumphantly hauled out of Barry during one very hectic week and positioned in the sidings at Bishops Lydeard. This trio was just part of the remarkable rescue of 213 locos from Barry, of which 150 have been restored over the years to working order.

While funds were available to purchase the Prairies, resources to restore them were not secured. The locos just sat in all weathers looking rather forlorn. Decisions were eventually made about their long-term future and No. 4561 was kept for restoration, 5521 was sold to a group on the Dean Forest Railway, and No. 5542 was purchased by an independent WSR-based group.

In 1979, No. 5542 became the property of the Locomotive 5542 Fund (and subsequently Locomotive 5542 Ltd) and restoration began in earnest in the open in 1980 at Bishops Lydeard. By this time the condition of the loco made it a sorry sight. The tanks, cab, boiler, wheels and frames were separated and many surfaces were painted with red lead paint to prevent further deterioration. The working parties sheltered from the rain in an old van body, and curious visitors put a few coins in a makeshift donation box – the loco's chimney!

No. 5542 had emerged from Swindon Works in the summer of 1928, only a decade after the end of the First World War. She spent a great deal of time based in the West Country, including at Taunton Shed where she would have been a regular sight on the branch-lines to Yeovil, Chard and Minehead. Her last posting was at Westbury, Wiltshire, from where she was withdrawn in 1961 to begin a 14-year sojourn in the salty air of Barry. As a Minehead regular she has a special place in the hearts of her shareholders and the wider West Somerset Railway community. Travelling behind her through the foothills of the Quantocks and along Blue Anchor Bay is one of the most truly authentic experiences on a heritage railway anywhere in Britain.

Early restoration was a slow process with significant funds being hard to come by. However a big funding programme was got underway, and having contacted many WSR shareholders for support, funds now started to accumulate at a much faster rate and progress could really begin in earnest. The frames and wheels were the first to receive attention at Bridgnorth, on the Severn Valley Railway. Gradually other parts followed and many of the missing parts were located or manufactured. In early 2001 an arrangement was made to move the partially-completed loco to Washford where work could continue towards completion much closer to home. The repaired boiler arrived in 2001 and the reassembly was completed in early 2002. A huge effort was made to finish the restoration work later that year and No. 5542's debut was at the WSR Autumn Gala on 3rd October that year. The loco's very first train was double heading with King Class No. 6024, something that could only have been dreamed about for most of the previous 23 years.

The other two Prairies were restored to life over many years. No. 4561 was originally built in 1924 and was restored at Minehead, steaming again in August 1989. After 10 years' use including starring in the film *The Land Girls*, she was retired for an overhaul of her boiler. For various reasons, including lack of funds, she has not yet returned to service but a renewed effort has started to get her back where she belongs, on the Minehead line.

No. 5521 headed off to the Dean Forest Railway in 1980 and she steamed again in 2007 after a three-year restoration at the Flour Mill loco works. In March that year she made a historic return to West Somerset as a special guest at the Spring Gala. She left Hull for a European tour the next year and was seen on railways in Poland and Hungary, including a stint hauling the *Venice Simplon-Orient-Express*. In 2013 No. 5521 was painted red as London Transport No. 150 as part of the celebrations of the 150th anniversary of the Metropolitan Line.

Apart from the Prairies, the West Somerset Railway story has featured many of those former rusting hulks from Barry scrapyard. In 1970 the Somerset & Dorset Circle rescued LMS 7F No. 53808 and in 1987 she joined the ranks of fully-restored locos on the WSR, continuing to give good service today. Her sister No. 53809 has spent a good amount of time on the WSR but has now gone off to pastures new. One of the biggest restoration projects on the line was Bulleid Pacific No. 34046 *Braunton,* which arrived in a very poor state from Brighton. After a lengthy overhaul she emerged from Williton shed in 2007 and fully re-launched in September 2008. *Braunton* is now fully certified for the main line and runs as former Battle of Britain class No. 34052 *Lord Dowding* in honour of the centenary of the Royal Air Force.

Dai Woodham waved good-bye to the last locomotive to leave his yard in 1990. GWR Prairie No. 5553 headed to the Dean Forest Railway and was eventually restored by the Waterman Trust, spending a lengthy time on the WSR. Dai died in 1994 but his legacy lives on, steaming around Britain as part of the £1 billion-a-year heritage railway industry that he created by accident rather than design.

Prairie tanks 5542 and 5521 stand coupled to a line of stock that appears to include the Peckett tanks at Bishops Lydeard. Peter Triggs

A Night to Remember

Steve Edge

It was a stormy December evening in the early 1980s when the former industrial steam locomotive Bagnall 0-6-0ST *Vulcan* set off for Minehead with the *Quantock Belle* dining train. In fine weather in daylight and with careful driving, *Vulcan* would just about manage the trip to Minehead and back, but the conditions were not kind. On this dark night with the other staff at Bishops Lydeard station, I watched the bright tail lamp as the train curved away out of sight and headed slowly up the bank towards Crowcombe, with at least four hours before the train returned.

After a while, fellow volunteer Norman Hawkes, with furrowed brow, spoke up and reckoned that the train might reach Minehead but what about the return trip, especially the steep climb from Blue Anchor to Washford. Could we help? A couple of hessian bags filled with fine sand were found in a corner of the goods shed, loaded into the back of my old estate car and then we were away into the night heading for Washford station.

On arrival, clad in high visibility vests, even though it was pitch black, and each with a sandbag on shoulder, we trudged along the track towards Blue Anchor. The steeper part of the climb from the coast ends just west of the accommodation bridge near Halscombe House, and that point was to be the start of our sanding operation. We debated how best to sand the rails, eventually deciding to treat just one rail to ensure a longer stretch of sanded metals. The sandbags became lighter as the sand trickled on to the wet railhead. If it had been windy it may well have blown away and our triumph would be short lived, but thankfully the dampness helped the sand to stick.

The sand ran out. We wondered what else could be done: maybe small pieces of ballast placed every six feet or so on the rail might be crushed by the locomotive's wheels and provide welcome grip. That's what we did for another 100 yards or so. Soon, we caught the sound of the Bagnall coming up the bank, very slowly, slipping with every exhaust

West Somerset Railway early days. 14th June 1981 sees Victor *arrive at Bishops Lydeard with a train from Minehead. The signal arm has been replaced, but the signal post (like the others at BL) was not recovered by BR even though the signal box closed 10 months before the line.* Peter Triggs

beat. A pause, then another beat, a slip and a pause. Clearly with the fireman shovelling sand under the leading wheels and the driver expertly managing the regulator, progress of sorts was being made. We started to walk back to Washford, listening to that painful struggle between an industrial locomotive and wet rails on a 1-in-65 gradient.

Eventually the glow of the headlamp and the open firebox were seen at the point where we had stopped placing small stones. Then there was the sound of popping stones and a slow regular exhaust beat. We stopped and stood well aside as the train, moving steadily at well under half walking pace, drew level. A cheery thanks from the footplate crew as they spotted we stone-placers. A few more yards and the wheels reached the sanded rail. What a difference! A stronger beat from the exhaust and the train started moving a little faster. Speed dropped when the train reached the end of the treated rail but with momentum and a more favourable gradient, the platform at Washford was reached followed by a welcome downhill run to Watchet.

Looking back at our unplanned and unofficial mission, it was perhaps not a wise move but, heck, we were so pleased to have helped get that train up the hill. Happily the loco and its crew made better progress on the longer climb to Crowcombe. Arrival at Bishops Lydeard was not overly late, and the *Quantock Belle* was met by two platform staff and the passengers thanked us for the trip. But few if any knew about our earlier efforts in the rain on Washford Bank.

It is sobering now to recall how the real possibility of a train not making a return journey was a constant consideration in those pioneering days!

A View From Washford

John Simms

By the time I started volunteering at the Somerset and Dorset Railway Museum Trust's base at Washford station, trains on the WSR were about to start running between Minehead and Bishops Lydeard and the scene at Washford had calmed down after a very hectic spell. The Trust had had to abandon its Radstock Project in 1975 and, having taken over a Washford site that had been reduced to a running line without sidings or goods yard structures, the volunteers had set about creating a restoration base and museum.

Prodigies had been achieved by the workforce with track laying, refurbishing the signal box to make a working exhibit where visitors could try their hand, and starting to show artefacts in a museum. The disruption hadn't been popular with all of the village residents, one of whom had applied for a household rates reduction because his house "now overlooked a scrapyard", but it had settled down and we and the West Somerset Railway were now accepted as part of life.

Not that I would have been much use in any of this: for me, a hammer is a thumb mutilation device and a paintbrush unwise, but I could talk and take money and give change, and so I became a museum and shop staffer. This was the public face of the Trust and also of the WSR, which as the line moved into the 1980s became distinctly interesting. The travails of trying to maintain reliable services with the then motive power will be found elsewhere in this book but suffice it to say timekeeping was erratic and too often appalling as machinery broke down. The honest answer to "What time is the next train?" would be "Your guess is as good as mine" but that would hardly give any impression of competence or professionalism. To compound matters, Washford didn't have a public phone and if the internal system was down, the only way to contact the rest of the railway was to use the payphone in the Washford Inn if it was open – those were the days when pubs closed in the afternoon and there were no mobile phones – or the phone box about a quarter-of-a-mile away. Often you reached the public phone just in time to hear the train whistle.

The lowest point at that time was the day I arrived to find a WSR notice saying that as there was only one diesel railcar available for service and no working steam engines, a revised timetable would be in use. I awaited the planned arrival of the first train, but that time duly passed with no sign of it. Contact was made with Minehead to reveal that the DMU had derailed and there would be no trains at all. I left a sign to this effect on display, locked up and went home.

Neither could we be unaware of the parlous financial state of the West Somerset Railway. Having been advised of the likely arrival of a Receiver, a cold and damp Sunday was spent getting certificates of ownership on to every item that the Trust owned. The crisis was averted but there were painful discussions in the coach that served as mess and accommodation as to where the Trust might end up next.

However, the dark days faded as the WSR turned the corner and there were plenty of lighter moments. As the 1980s progressed, the line-up of Washford Museum staff came to include Roy and Mona Pitman and their black labrador Sooty. Like all of our canine chums, Sooty had definite ideas of what was entailed in being a dutiful dog, and at Washford he regarded it as important that when he heard a train whistle he should climb on a platform barrow and bark. One warm Saturday I was on duty but Mona had come down to do some dusting and polishing and she was in the back room with Sooty asleep on the floor. We had passengers enjoying the canopy shade and all was peaceful until the whistle sounded. Cue Sooty tearing out of the door barking. This upset a passenger's waiting dog that added to the decibel count and soon acquired a descant from a little girl in a pushchair who understandably took fright. By the time Mona, apologies in full flow, was pulling the miscreant back by his collar the tranquil wayside station was bedlam and to round it all off, an enthusiast was at the front of the train with recording kit, looking rather bemused. Then there was the day Sooty decided to

This picture is from a slide that was just labelled "Washford Early Days". The SDRT have clearly started their work to develop the site and the steam crane is in place. Thought to be around 1976, when John Simms began volunteering in 1978 much had been done to transform this scene for the better. Tony Yardley

board the train and see his friend the guard. He was returned on the next train from Watchet with a note saying that although he had forgotten his staff ID, no fare would be charged!

Joining the world of telephones was a blessing, but its busiest day was after a school's visit to the museum. The teenagers were interested and easy to deal with but, before going on their way, it was packed lunch time. That was the signal for two of the girls to go topless sun bathing on the saddle tank of the out-of-use Peckett *Kilmersdon* at the Minehead end of the headshunt, just in time for the lunchtime train from Minehead to roll in. The story was soon up and down the length of the line, no doubt losing nothing in the telling, and the humourists had a field day. Today *Kilmersdon* is restored and often appears in company with SDJR coach No. 4.

The real heart and soul of the Trust then was Secretary Mike Palmer, who had eternal patience and dedication but was affectionately known as "Captain Mainwaring", and there were some Walmington-on-Sea style episodes. The arrival of the unrestored coach body of No.4 was one such. Mike arranged delivery for a summer Saturday, blissfully regardless of the fact that in those days it was changeover day for Butlins at Minehead. By the time the low-loader had manoeuvered on to site, traffic was stacked up for miles around while Mike was obliviously directing in the middle of the road. But without Mike and his ability to deal with some forceful and clashing personalities, much less would have been done.

Another steam engine on site in the early days was 7F No. 88/53808, which a declining number of hardy souls were trying to return to steaming condition in the open air and later in a makeshift "tent". The current restoration shed was then some time in the future. Progress on the loco at Radstock had been steady, but at Washford it slowed and the contract for overhaul was therefore placed with the loco department at Minehead. This was an act of faith as Minehead had little overhaul experience, but it paid off when the big 2-8-0 entered traffic in 1987. It was part of the changing fortunes of the WSR and over the last three decades has continued its career of plodding along steadily with whatever the railway requires it to pull.

At Washford the signal box from Burnham-on-Sea had arrived as had the goods office hut from Wells, which was extended to give a better quality of volunteer accommodation and allowed the former coach to move to Williton. An indoor toilet in the same building as the bedrooms was a major step in itself, although that involved a team of three working for a week in a heatwave to install a septic tank. The eventual mutiny by two thirds of that workforce in favour of an evening in the pub rather than continuing with the task enlivened my arrival in the mess room after I'd locked up the Museum. And yes, I did join all three in licensed premises where good humour was resumed.

Other bits of kit drew visitors, notably an enthusiast for Ruston Bucyrus diggers from the North of England who travelled down and paid his admission to see Tick Tick, our faithful hole-creator standing in the yard. You could never predict what the day could bring.

In 1994 I joined the paid staff of the WSR at Minehead and my staffing days at Washford ended. It was a decade-and-a-half I very fondly remember, as the Trust achieved a great deal and the passing trains became more reliable and steadily fuller.

The restoration of Crowcombe. Above, the sight that greeted Walter Harris and Mike Hodge when they took on stewardship of the station. Virtually derelict and rapidly being reclaimed by nature. Alan Turner

Below, 31 March 2012 represented 150 years since the railway opened in 1862. Here we see a fully restored Crowcombe dressed in flags to mark the event. The photo provides a contrast to the one above showing how much work has been done. Ian Coleby

Thirty-Four Years at Crowcombe

Walter Harris and Mike Hodge

Walter writes: Railways are in my blood. My father was a Great Western signalman who began his career at Princetown, on Dartmoor. Later he moved to Carne Point on the Fowey branch, and then was promoted to signalman on the mainline at Royal Albert Bridge signal box, controlling the main line from Devon into Cornwall. I spent many hours in this box helping my father until I began my teaching career at St Luke's College in Exeter, which led to me taking a job at Cannington, near Bridgwater.

My good friend Mike Hodge was an even greater railway enthusiast than me and as soon as he heard that the Minehead branch was to be closed, he became involved in the plans to reopen it. He attended the inaugural meeting in 1971. We joined the WSRA in 1972 and started helping on the railway. Michael and I were part of the early working parties at Bishops Lydeard in 1975, clearing the track and platforms. Those early days were pioneering ones and we became very enthusiastic about the chances of reopening the line. I was a schoolmaster and Mike was an insurance man and the break from our day jobs to do much more manual labour in the open air was very welcome.

One day, while we were working at Bishops Lydeard we were approached by Hein Burger, who asked if we would like to take on a station! At that time, either Dunster or Crowcombe was available and we quickly chose Crowcombe for the simple reason that it was the closest to Cannington where we both lived. We became joint stationmasters - we had no idea how much work we had let ourselves in for!

Like all of the other stations on the branch, Crowcombe had been virtually abandoned by British Rail and was now almost derelict. The scene that greeted us when we made our first visit was disheartening, but both Mike and I and our wives enjoy a challenge, and this was a very big one. The station was in an awful state. The windows had been boarded up and the glass was all smashed. The doors needed repairing. There was no water or electricity. There was no floor in the booking office or waiting room as the floorboards had been taken as firewood. The booking office chimney was leaning at a drunken angle. The signal box had burned down in 1966 and all the signalling had been removed. Our first jobs involved much tidying up and endless hours weeding (with pickaxes!), bramble removal and having great bonfires to get rid of the rubbish.

We'd decided at an early stage that we'd be responsible not only for the station but also the banks between the two road bridges. One day we found Harold Blackmore and John Sumbler working in the station approach road clearing the ditches. We couldn't understand why they were doing this and I remember Harold's reply to my question very clearly: “Boy, that's railway.” By this we knew that this approach road would also be for the railway to keep clear, and this has occupied a lot of effort and money over the years. Clearing the railway of weeds and brambles was a marathon job and even working with pickaxes it was only possible to clear a few yards at a time. Those first few years before regular trains began were quiet. We might see an engineering or works train and we both remember very well the 1975 Directors' Special – in reality, one of the Bagnall tanks and a brake van – driven by Harry Lee, with Cedric Dunmall as fireman, passing through.

An early job was to take care of the dangerously leaning chimney. We asked Brian Stainton to have a look. Some readers will know that the water supply to the Stationmaster's house (now Puff Cottage) was from a hand pump on Crowcombe station platform. It had been the junior porter's job to fill a header tank in the loft of Puff Cottage using this pump on the platform. So that he knew when it was full and to avoid flooding the house, the overflow pipe came all the way back down the lane and up the chimney at the station. When water came out of this pipe, the porter stopped pumping! When Brian examined the chimney, he found that the pipe was pretty much the only thing holding up the chimney and we had to have it taken down. Brian took off the first few courses of brick and the job was completed by my wife, Margaret. Eventually we uncovered the well on the platform. It proved to be 54ft deep with 27ft of water in it. We never found the pump.

When regular trains restarted, we had been joint stationmasters for six years. Five years later, in 1984, the Board was considering closing the line beyond Williton to save money. We had a visit from John Brown, of the tourism authority, Doug Hill, then General Manager, and two West Somerset councillors. I am not sorry to say that they got short shrift from us and we reminded them that the main terminus for passengers on the railway was Bishops Lydeard and if that section was closed, then it would be the end of the railway. They clearly listened to us and the whole line was kept. Several years later we had a visit from a Voyager unit from the main line and we were asked to travel on the train. We were embarrassed when David Morgan introduced us as "the men who saved the railway.”

Right from the start I was keen to reinstate the signal box and passing loop, which had been removed by British Rail. This could only be a distant dream while more important work had to be undertaken and we could find a suitable signalbox. We then had a phone call from Chris van den Arend at Williton, who told us that he had met a Mr Thomas who’d seen a redundant signalbox at the tinplate works at Ebbw Vale. This box had been Ebbw Vale Sidings South and luckily was exactly the size we needed. We went to see the Steelmaster to ask if we could have it. We were told that they were keen for it to be taken away and had offered it to several heritage railways, but none had been able to fund its removal. He told us that we could have it for nothing provided that we took it away and cleared the site.

After making detailed arrangements we organised two cranes, and Steve Harris masterminded the lifting of the wooden top of the box using two RSJs on to a low-loader. We agreed to remove the bricks and we also had to take off the slates to avoid them flying down the M5! Eventually, after several heart stopping moments, the wooden top arrived at a temporary home on the loading dock at Crowcombe station. Volunteers dug out the footings of the old signal box and we built a suitable new brick base to fit the rescued top. The locking room windows were also from Ebbw Vale, but we had to have new window frames made to replace those that had rotted.

We were grateful to the Taunton group of the Great Western Society who erected the new signals that were needed. The posts were cast in moulds with concrete and were completed with "new" Great Western-style arms and fittings. Although the work took several years, we were most satisfied when the signalbox and new passing loop were opened and the station could return to being a passing place again.

When the station was opened in 1862, it was called Crowcombe Heathfield. In 1889 the name was shortened to Crowcombe to avoid confusion with Heathfield, in Devon. This might have solved that problem but created a new one. The station is actually more than two miles from the village, and this confused visitors. More than one delivery lorry got lost in Crowcombe village looking for the station. One day a vicar climbed off an up train there. I was on duty and he told me that he wanted to “just nip over to the churchyard to visit the graves of some relatives.” I reminded him that the village was an hour’s walk away. He was very unimpressed and mentioned the Trade’s Descriptions Act, and this led us to apply to revert the name to the 1862 original – not very easy as we had to have new running-in boards made, but we are glad we did it and it has avoided more lost vicars!

Along with the signalbox, the down platform shelter was lost when the station was closed by BR. While on his travels to Bristol, Mike Hodge spotted a redundant bow-roofed GW hut in a quarry alongside the line at Flax Bourton. Enquiries revealed that it belonged to Sir Jack Hobbs, who owned the quarries, and he gave us permission to remove the hut to Crowcombe. Initially sited on the down platform, the hut was never really suited to a waiting shelter and we later moved it to a new position on the bank and commissioned a replica GW shelter in its place.

It would be remiss of me not to mention Ted Cubbit, who joined us in 1978 and made a massive contribution to the station as a master electrician and the person in charge on a Saturday. Thank you to him, his wife Jan, and all of our loyal volunteers over the years. Michael and I retired as joint stationmasters in 2009 after 34 years at Crowcombe, having rescued the station from a derelict wreck to become a showpiece on the West Somerset Railway.

The team are seen erecting a station lamp at Crowcombe. Walter Harris

Memories of a WSR Blue Anchor Signalman

Richard Derry

It all started with the death of my father in October, 1984, when I decided to take a more active part in a preserved railway. The nearest to my then home was the Mid-Hants, which saw me start as a volunteer as part of the station staff at Medstead and Four Marks station. Within 18 months I had been passed to operate Alresford, Ropley and Medstead signal boxes. Meanwhile my mother had started taking holidays at Minehead, where I had some training turns on the footplate on the West Somerset Railway. I quickly realised I was not cut out to be a fireman so I decided to train as a signalman, starting at Blue Anchor signal box. The WSR was operating with two separate train staffs to Minehead, one allowing access to the main platform and the second to the bay.

With a basic timetable of four trains each way, in the afternoon we had two down trains in a row and when the second arrived at Blue Anchor, both train staffs were at Minehead. So one of the staffs had to be returned by road, and more than once someone had to use the bus service to deliver it. Once when I was operating BA, I had no train staff and had to hold the DMU in the platform, with David Rouse driving. Until it had been delivered, I hid in the box to escape his wrath! The messenger and I had to sign and counter-sign the Train Register to say which train staff had been delivered and at what time. Then I could hand the staff to the driver, ensure the points had been set, operate the gates and pull off No. 3 and No. 4 signal levers to allow the DMU to depart. Only after it had cleared the points, the signals had been returned to danger and the gates dealt with could peace once more return.

At this time I was keeping a detailed diary of my railway activities with the intention of getting it published. This was not a financial success and I have since turned to other railway topics to get my name in print. I stopped my diary in 2000. What follows are extracts:

9/6/87: I was on a week's holiday and Steve Martin had arranged a training day in Blue Anchor box. I arrived at 10.06 and although the lady signalman, Andy (short for

The interior of Blue Anchor box in 1976. On the left is Nick Jones, then operating superintendent. The gentleman on the right is unknown though thought to be an ex-BR signalman who worked the box before 1971 and visited again during the first summer. Allan Stanistreet

Alexandra), wasn't expecting a trainee, she made me welcome. I watched her operate the frame for the 10.30 up train and was then allowed to have a go. Though a passed signalman on the MHR, I hadn't operated level crossing gates before, so first having checked there was no traffic in sight – the box giving a good view of the B3191 Carhampton to Watchet road – Andy instructed me and I felt at home except when I put No. 13 up starter back instead of No. 14 up home! Of interest to me was a visit to the locking room, which you have to be careful entering and leaving, as the door is right next to the road.

25/8/87: I'd driven down on the night before (Friday) with a friend. In atrocious weather we arrived beside the Home Counties camping coach in Blue Anchor station yard. It took 10 minutes in the dark to find the main electricity switch and then get our gear and settle down for what was a cramped and stuffy night. But I didn't have far to go in the morning to the signal box. I noted Andy walking along the line from the Minehead direction, and she was soon in the box. We both signed on in the Train Register then she was off down the box steps to sell tickets, leaving me to open section with Williton box which also had a trainee on duty – Chris Austin. A number of times No. 6 points wouldn't go right across, and No. 2 down home wouldn't come off. The problems were solved by a visit by Roger Woodham, then a member of the S & T gang, his ministrations soon sorting them out. Another problem was with one of the stops that come up in the middle of the road when the gates are opened against the traffic It wouldn't go back down and had to be "persuaded" to do so by Roger clearing the muck accumulated in the drain beside it. Once the day was finished, we had a meal in the Blue Anchor Hotel and when we got back to the Camping Coach we found four more people in residence although we had been promised the coach to ourselves.

16/9/87: I was down on holiday with family and friends and staying in Minehead. When I arrived at 09.50, Andy was already on duty. The 10.15 from Minehead was late as extra carriages had to be added, and it left at 10.31. By now I was used to operating the gates and it was nice to see steam coming from Minehead, giving plenty of time to get them open. Once the 16.30 from MD had cleared and we closed the section I gave a lift to Andy to Dunster Station as she wanted to go blackberry-picking. I wanted to thank her for the excellent job she made of polishing the original buttons on my waistcoat...which is far too small for me now!

16/9/87: Again a late departure from BL for the 10.35 because the rostered guard hadn't turned up, and it eventually departed at 10.52. At lunchtime, Andy showed me how to issue tickets, as then the Signalman was also expected to act as booking clerk. In fact, this was a duty I never carried out as often people arrived just before the actual train they wanted!

7/5/88: During the day the Minehead-end points, operated by levers Nos. 10 and 11, had jammed and according to the Train Register had been disconnected, locked and all up trains had been using the down platform, hand-signalled through. I watched Roger deal with the points and rodding. It was dark by now and I heard a church clock strike 10. How many people, let alone enthusiasts, realise the work that goes on after dark to keep preserved lines running? But that wasn't the end of a disturbed night in the Camping Coach, for I was awoken by a brilliant flash. A car had somehow overturned coming down the hill from Carhampton, skidded on its roof, hitting one of the electricity poles. This was followed by the arrival of fire, police and ambulance, plus an electricity maintenance crew.

8/5/88: My signalling test was scheduled for the afternoon, taken by Mr. Harwood, a BR inspector, but I wasn't able to have a free morning as the first down train arrived with the BA-WN train staff when it should have had a ticket, so I had to take it back by road to Williton. Then I drove back to BA, changed into my uniform and took up position in and around the station. Mr. Harwood arrived, Andy signed off, and I was put through rules, regulations and frame test. The questions included what to do if a train staff is lost; what happens if all communications between boxes fail; what is a clearing point; what do you do with trains approaching from both directions; what class of train do you use to examine the track, and many more. I passed with a reminder to consult the WSR Rules and Regulations, but I now required a medical.

4/7/88: This was my first solo turn in BA signal box. I had a number of visitors, including a gentleman who had worked for 50 years at Swindon Works where he was foreman in the parts department. When he retired in 1979 the place contained 30,000 parts worth £10 million. He looked at several of the signalling and PW items and quoted serial number, measurements etc. Once I had signed off it was drive into Taunton and at the station I noted No. 08849 acting as pilot, No. 47049 on a freight train, and a brace of 37s on an oil train. How times have changed.

I remained a passed BA signalman for nearly 30 years and later passed to operate Williton and Crowcombe Heathfield signal boxes. One of my favourite turns was a late summer Saturday evening when No.92220 *Evening Star* was down for one season. On a damp evening it was very atmospheric as the 2-10-0 passed through BA at a sedate speed and then opened up for Washford Bank. I still volunteer on Dunster station along with a good crowd of people, though I don't neglect BA where I help Santa every Christmas.

Reflections of a WSRA Secretary

Brian Crudge

Soon after coming to Taunton in January 1971 I joined the WSRA and was a member of the first working party at Bishops Lydeard. During one of our early visits, someone whom we had never seen before turned up in a Morris Traveller which appeared to contain all his worldly possessions. We wondered if he was an individual who had caused grief and expense by objecting to the Light Railway Order because he claimed trains would disrupt his holiday home and wildlife, and perhaps was now working "under cover" to discover what was going on. However, this was actually Tom Waller, who was very friendly and knew quite a bit about the WSR as well as being a volunteer on the Kent and East Sussex Railway. His conversations about what was involved in restoring that railway gave us the encouragement to continue our work, and he became quite involved in many ways over the years.

In 1974, John Pearce talked me into becoming Secretary. At that time differences between the Association and Company continued to simmer and I undertook some meetings with then-Company Chairman Douglas Fear to try to find a resolution. The outcome was liaison meetings between WSRA and Company representatives, which worked well and were resurrected in similar format during my second secretarial stint.

John became a friend for more than 40 years. I was Deputy Head at Bishop Henderson School, Taunton, where both our sons were in the same class, and also through the early Association working parties. Perhaps not many knew that beneath his many attributes, while not a church-goer, John was a man of great faith. We had many discussions on his lively life-style, while for a number of years he had me "booked" for his funeral service! He used to say: "All church people should be like you," then if I made a doubtful utterance when things weren't going well, "Now 'vicar', you shouldn't say things like that!"

John was known even on a good day to have a short fuse, which encouraged some to wind him up, often successfully. I found I could get away with whatever I said to him, apart from once when we were discussing the Bishops Lydeard shop at a committee meeting and I made a simple suggestion about its organisation. That led to the immediate retort: "I don't tell you how to teach, so don't tell me how to run 'my' shop!" A discussion at another committee meeting also sticks in my mind: we met in John's house but he as Chairman wasn't getting his own way. Suddenly he stood up saying he'd had enough and went out, slamming the door behind him. We sat there wondering what to do when we heard tapping and his voice calling: "Please can you unlock the door and let me back in." He had accidentally shut himself out, and on his return we continued as though nothing had happened. John lived for the West Somerset Railway and, though they must have been frustrated at times, he had the continued support of his family. Many miss him greatly although we have splendid memories of a great friend and lively character.

On reopening day in 1976 I organised the second train, the "WSRA Special" hauled by No. 6412, recently purchased by the Association. At that time most committee members were assisting with the running of the Company so I was left keeping committee matters going almost by myself, and I also organised the Taunton area meetings. The following year I resigned as Secretary and was rewarded with WSRA honorary life membership.

Over the years I have continued to keep in touch, not least as main commentator for the WSRA's Great Western 150 celebration parade and the first 29 or 30 Steam and Vintage Rallies. John hauled me back again as Secretary in 1991 and I carried on doing the job that nobody else wanted, assisting setting up the WSRA as a company limited by guarantee and registered charity. Then the Association discovered No. 34046 *Braunton*, still in unrestored state, was for sale and agreed its purchase. Consideration was given to beginning its restoration once overhaul of No. 6412 was completed, but events overtook us. Out of the blue came a letter from Jeremy Hosking offering to buy *Braunton* and fund its overhaul. Chairman Humphrey Davies was authorised to negotiate an agreement under which Mr Hosking would pay the WSRA £250,000 for the purchase of *Braunton* and its restoration to main line standard – this cost considerably more, but he fully covered it. The Association was to have use of the locomotive for 10 years, with an option for further use after overhaul.

Gareth Winter was appointed Mechanical Engineer, with his salary covered by the *Braunton* fund. The remains of No. 34053 *Sir Keith Park* – mainly wheels and boiler – plus a van-load of spares were purchased to assist the restoration of *Braunton*. Inevitably the overhaul took longer than expected, and as soon as it was ready to run, a crack was found in a driving axle. This was remedied through a part-exchange with the group restoring No. 35011 *General Steam Navigation*, so in due course *Braunton* was ready for service and has been an excellent performer.

Jeremy Hosking made regular visits and was always supportive of the restoration team's efforts, having great regard for Gareth Winter. This was his first locomotive purchase but from little acorns, mighty oaks do grow! And I must recount a story that his wife assumed he had purchased a model *Braunton* and he had to explain carefully that it was in fact a full-size loco. However, considering what he has achieved since, she must have been quite happy about it!

Fireman – For Just One Day…

Barney Forsdike

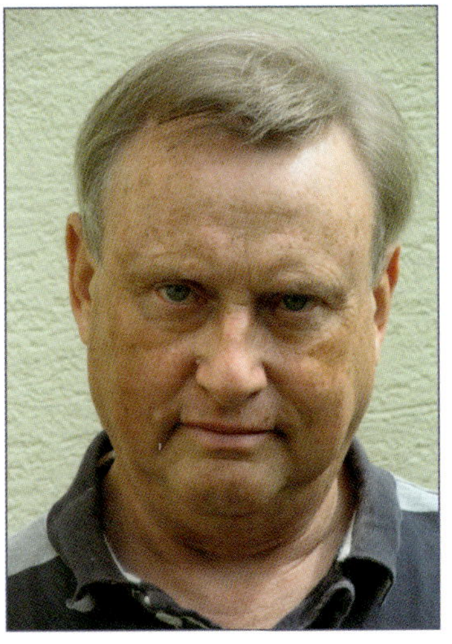

It must have been in the late 1970s that I had my one and only firing turn. No, it wasn't part of some pre-booked footplate experience course but a last-minute job done out of sheer necessity. I had already been a regular weekend volunteer at Minehead for a couple of years, mainly doing booking office duties, acting as Travelling Ticket Inspector or helping with publicity, and I had been training as a guard. At the time, the railway was run by quite a small group of people, and multi-tasking was the order of the day.

On the morning in question I was walking along Minehead platform when David Rouse approached me and asked if I was busy. David was a regular steam driver on the WSR, a preservation pioneer who had worked on the revived Ffestiniog Railway in its earliest days, and had been involved in saving a number of locomotives, culminating in his role as custodian of Collett 22XX class No. 3205. An ex-Western Region engineman, he'd also accumulated an encyclopaedic knowledge of GWR rolling stock and was a fascinating man to talk to as well as being, quite simply, a very nice guy.

David told me that his booked fireman was a "no show" and that there was literally nobody else but me available. At the time I had scarcely ever been in a loco cab, had no prior experience, and never aspired to working on the footplate either then or since! But there seemed to be a straight choice: either I would do it or the train would be cancelled. Of course I agreed and went off to put on my oldest working clothes.

The rostered loco that day was Bagnall tank No. 2996 *Victor,* which David had already prepped and attached to four Mark 1 coaches in the main platform. The two Bagnalls owned by the company were powerful and sophisticated shunting locos, fitted with all the latest 1950s mod cons. However, they were quite ill-suited to what was asked of them on the WSR and their short wheelbases made for a jerky ride if you were at the front of the train and trying to drink your tea! We were booked to do two round trips from Minehead to Williton, a total of 40 miles with a steep hill in between.

David had already ensured that the fire was good to go and that the tank was full, so my baptism of fire commenced with tuition on how to work the injectors. These devices can at times be temperamental, but I must have been lucky that day as I experienced no problems with them after following David's instructions. Putting coal on the fire was more tricky – I often failed completely to get it in the right part of the firebox! David had to tell me more than once not to over-fire – I think I was petrified that we would run short of steam and get stuck going up to Washford! I already knew from my guard's training how to safely couple and uncouple the loco from the train, and that I would need to relay the guard's "right away" signal to David. Watering the loco does not stick in my mind, so I suspect that I must have been left in charge of the valve. I do not have a head for heights and would have remembered clambering up that saddle to the tank filler. Nor do I have any memory of exchanging the wooden train staffs at speed in the up direction at Blue Anchor, so probably this was done when our shortish train came to rest in the platform.

This story would probably have been more interesting if we had run out of steam or water, or I had done something foolish to delay the train. But neither of those things happened and the day seemed to be passing uneventfully – we were almost "right time" on each of our arrivals at Williton and Minehead. Although tired because of the unaccustomed physical exertion, I even began to enjoy myself on the homeward trip in the afternoon. I'm sure that this was entirely down to David's expert tuition and easy-going manner.

There was, however, to be a sting in the tail. We had got our train safely back to Minehead on the final trip and were taking the loco on shed at a very gentle pace when I must have let my concentration lapse for just a second. I was probably quietly contemplating how easy this firing business was and wondering why was there so much mystique attached to it. Without warning, David suddenly pushed down on the regulator on his side of the cab, and the other end shot up and hit me on the forehead just over my eye socket. No blood was spilled, but I soon developed a lovely "shiner" as a souvenir of my day on the footplate. I should explain to those without long memories that, as built, the Bagnalls could be driven from either side of the cab and were therefore fitted with a centrally-pivoted and rather lengthy regulator, accessible from both sides. When I got back to my paid job the following day, I tried jokingly to convince my friends that I had been on the wrong end of someone's fist in a Minehead pub, but I think they knew me too well to believe that!

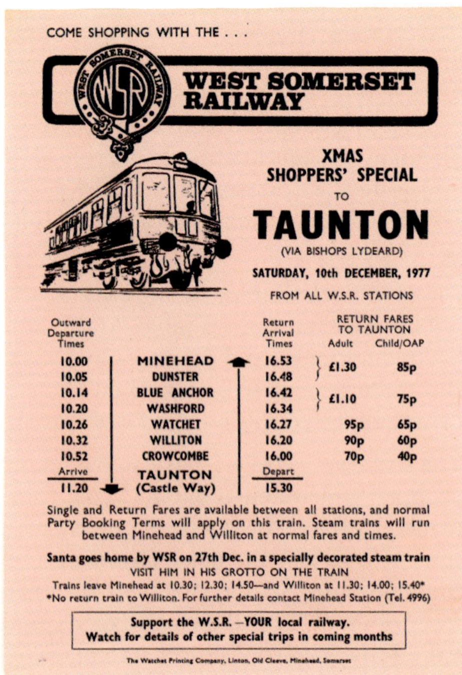
COME SHOPPING WITH THE . . .

WEST SOMERSET RAILWAY

XMAS SHOPPERS' SPECIAL TO

TAUNTON

(VIA BISHOPS LYDEARD)

SATURDAY, 10th DECEMBER, 1977

FROM ALL W.S.R. STATIONS

Outward Departure Times		Return Arrival Times	Return Fares to Taunton Adult	Child/OAP
10.00	MINEHEAD	16.53	£1.30	85p
10.05	DUNSTER	16.48		
10.14	BLUE ANCHOR	16.42	£1.10	75p
10.20	WASHFORD	16.34		
10.26	WATCHET	16.27	95p	65p
10.32	WILLITON	16.20	90p	60p
10.52	CROWCOMBE	16.00	70p	40p
Arrive 11.20	TAUNTON (Castle Way)	Depart 15.30		

Single and Return Fares are available between all stations, and normal Party Booking Terms will apply on this train. Steam trains will run between Minehead and Williton at normal fares and times.

Santa goes home by WSR on 27th Dec. in a specially decorated steam train

VISIT HIM IN HIS GROTTO ON THE TRAIN

Trains leave Minehead at 10.30; 12.30; 14.50—and Williton at 11.30; 14.00; 15.40*

*No return train to Williton. For further details contact Minehead Station (Tel. 4996)

Support the W.S.R. —YOUR local railway.
Watch for details of other special trips in coming months

The Watchet Printing Company, Linton, Old Cleeve, Minehead, Somerset

Twelve Years in the Life of a General Manager

Doug Hill

Doug Hill describes his experiences – and some unwelcome surprises – a couple of years after trains began operating between Minehead and Bishops Lydeard

I became General Manager of the West Somerset Railway in 1981 and sadly found equipment in a very poor condition and morale among both paid and voluntary staff at a low level. This was mostly caused by having three General Managers (including me) in three years, while the company did not possess sufficient experience of operating a heritage railway, having introduced services running the full length of the line without sufficient finance, equipment or staffing.

Pannier tank No. 6412 wasn't available because it needed an overhaul, *Victor* and *Vulcan* weren't satisfactory for running passenger services of 20 miles, and in any case were in poor condition, particularly the boiler tubes. When one engine was repaired, the other then needed urgent attention ad infinitum. For instance, one evening the dining train came in to Minehead and the driver asked me to look at the engine. I looked into the firebox and found water cascading down from leaking tubes like a waterfall. Needless to say, I failed the engine and the diners had to go back to Bishops Lydeard by one of the DMUs. These units were in a bad state, particularly their engines. When they were in use they had to be topped up with water, generally at each station, due to overheating.

Problems occurred frequently, such as derailments at Bishops Lydeard, Williton and Minehead. The cause of the Minehead incident was the fact that the signal box had not been commissioned yet and the lines into the main and bay platform were controlled from the ground frame. A combination of errors meant that the points had not been changed and when the train set off with its passengers from the bay platform it was promptly derailed by the trap point. I held an enquiry and took action, but as it was a reportable incident the Railway Inspectorate also held an enquiry, although luckily I had already taken the steps necessary to prevent it happening again.

Doniford Beach Halt was opened on 27 June 1987 to serve the holiday camp built on the site of the former Doniford army base. The platform is built from concrete panels recovered from Montacute on the closed branch line from Durston to Yeovil Pen Mill and the pagoda shelter was recovered from Cove Halt on the Exe Valley Branch. Allan Stanistreet

One day a steam engine struck a car on a level crossing, and the driver of another steam engine had to drop his fire at Crowcombe due to the low water level in the locomotive's tanks. Such problems meant we often had to run bus services, with one of our own staff driving. So steps had to be taken to make a foundation for the service we have now. The first was to reduce staff costs, and I was thus made redundant – and while serving my three months' notice I had to make others redundant until we were down to three paid staff plus myself. I also had set about planning the organisation of a railway which would only have three staff, and devising a timetable within our resources. Needless to say, three months weren't enough to correct all that needed changing, and my redundancy was rescinded.

Hein Burger became the voluntary Assistant General Manager and without his support plus Peter Rivett, Managing Director, and my wife, I could not have carried on. When David Morgan became Chairman, stability was assured. A period of consolidation then followed with the support of the West Somerset Railway Association and the Diesel and Electric Group. We also encouraged and obtained support from local industry including the then Taunton Cider Company and Hinkley Point Power Station. Costs were strictly controlled, and for a while to reduce our fuel bills we ran steam to Williton and DMUs from there to Bishops Lydeard.

Things also happened which were outside our control. One day as I walked out of my office, I couldn't see the track because the high tide and rough sea had completely flooded the station area, including the workshops. Another time I had to decide whether to take rolling stock to Bishop Lydeard when coastal erosion threatened to wash the track away at Watchet. Then there was the time I walked out of my office, shut the door, then heard a big bang inside. I found the cause was the ceiling falling in! Life was certainly hectic – no days off, no holidays and every chance of being called out in the middle of the night because the Minehead shop had been burgled or, once, because the crossing gates at Blue Anchor had been smashed by vandals on a hijacked permanent way trolley.

I remember I once drove the Class 07 diesel with a works train from Minehead and gradually lost momentum with brake problems until on the crossing the brakes came on full, with the works train straddling the crossing and causing hold-ups. Then one Sunday both steam engines failed so I had to take the train to Bishops Lydeard with the same small diesel locomotive and was met with dismay by the *Quantock Belle* staff whose coach was to be added to the train. It was a tough trip up Crowcombe bank, with both the oil and water temperature warning lamps coming on, but we made it and I knew the engine would cool down on the descent from Crowcombe.

Due to lack of experience of staff in the locomotive department the Railway Inspectorate insisted that I should be responsible for boilers. On one official visit Major Olver complained that I was General Manager, Responsible Person for boilers and also in control of operations. What he didn't know that I also guarded a train on the same day after he had left! Our equipment and the experience of staff improved until we could run a Sunset Special to Blue Anchor once a week as well as a Jazz Train one evening a week, which included the siphon van where customers could dance. Plenty of drink was consumed and one evening all supplies had gone by the time we reached Washford so that we had to restock from the public house beside the station. One winter when the snow was deep, Bill Monteith slept at Minehead station to keep an engine in steam in case it was wanted. It was, and we took trips with passengers and mail through to Bishops Lydeard, and hay to farmers who couldn't get it to their animals any other way.

The next step forward for the line was the restoration of the former S&DJR engine No. 53808, which attracted volunteers to work for us as well as increasing passenger numbers. As our

In 1981 the Great Western Society Taunton Branch celebrated their tenth anniversary. On 14th June, they chartered the .Quantock Pullman *for a special outing. Here, members of the society with David Hartland (as Brunel) are seen at Minehead with* Victor *ready for the return to Bishops Lydeard.* Peter Triggs

equipment and staffing levels improved, we were able to build Doniford Halt using job-creation people who also rebuilt the booking office at Minehead with materials obtained from Cardiff station. Walter Harris rescued the wooden top of a signal box from South Wales, and this was used in the rebuilding of Crowcombe signal box. Minehead signal box was recommissioned and later the level crossing was installed. We hired several tank engines at times from the Great Western Society of Didcot, and as the railway was now on a sound footing, we were able to hire in *City of Truro*, the first steam loco to achieve 100mph and which I had helped repair when I worked at Swindon works, as well as *Evening Star*, the last steam engine built for British Rail. Then came another connection to my home town when, after much negotiation and assistance, we were able to obtain half of a listed building from the old railway works and erect at Williton what is now known as the Swindon Shed. Because of my experience there I was asked by the Heritage Railway Association to help salvage various items before the huge old GWR complex was finally closed. In its last week I went with some WSR staff to meet a number of old mates still there. Everyone wanted to be helpful in offering components and tools, and with these we filled a lorry.

During my time on the WSR I saw a loss of £24,000 in my first year improve to a profit when I finished of £14,000. It was a tough period, but one I would not have wanted to miss and it certainly proved that problems need both hard work and the right decisions.

Target Taunton

Chris Austin, OBE

Chris has been a WSR volunteer since 1987. He is the Chairman of the WSSRT (from 2009) and was PLC Chairman from 2003 to 2008.

When the preservation of the Taunton-Minehead line was first proposed, it was on the basis of running a regular public service during the week using a diesel railcar, with steam trains at weekends. Freight was also a possibility. Following the formation of the company, initial discussions went well. After the first stage of the railway between Minehead and Blue Anchor reopened, a running powers agreement with BR was signed on 3rd May 1976 which would allow WSR trains to run into Taunton.

Subsequent extensions of the railway took it to Williton later in 1976 and to Stogumber in 1977. The following year, discussions with the National Union of Railwaymen (now the Rail, Maritime & Transport Workers' Union) were proposed but rebuffed by the union. Their opposition to trains running through to Taunton was based on the belief that this would threaten the jobs of their members who were employed by Western National to drive the buses between Taunton and Minehead.

In 1979 the line had been restored as far as the limit of WSR ownership at milepost 165¼ at Norton Fitzwarren. In April, the local manager for Western National at Taunton wrote to the WSR to advise that the shuttle buses planned between Bishops Lydeard and Taunton had been "blacked" by the union, which had instructed signalmen not to handle the rolling stock bought by the WSR and assembled in the yard at Taunton. This action by the Taunton branch of the union had been endorsed by its regional council and by the national executive. The WSR Chairman, David Morgan, had asked the Trades Union Congress to intervene, but even they had been powerless to change the union's decision.

Experience with operation of the WSR had caused the Board to abandon the original

Early optimism is demonstrated in this poster from 1974. The prospect of trains to Taunton was then the key and at the modest cost of just £65,000. WSSRT Collection

aspiration of running trains all through the year in 1978, and the NUR's intransigence meant that the public service to Taunton could not be introduced. From this point onwards it was evident that the WSR would not be able to offer a regular public service but would develop as a heritage railway, as had every other preserved line in Britain.

Track rationalisation and the introduction of the Exeter Panel Signal Box in 1985 swept away the signal boxes in the Taunton area, along with the signalmen who had refused to handle WSR trains, but it also severed the WSR from the main line, with the abolition of the Up Relief line, and the conversion of part of it to a siding for Taunton Cider. However this was not the end of the story, and five years later an agreement with

BR and Taunton Cider allowed the railway to be reconnected through the cider company's siding, and protection was provided by a sleeper barrier padlocked to the rails and two stop boards. On 16th June 1990 the first through train ran over the new connection, a Minehead to Manchester excursion. I was the hand-signalman appointed to accept the return train to the WSR and had the job of unlocking the sleeper barrier, handing the BL-NF train staff to the driver and flagging him past the stopboard.

In 1991 I joined the board at the same time as John Donner, who lived at Stogumber and had recently retired from Lloyds of London. John was keen to see if more could be done to provide a regular service over the link to Taunton and the Board delegated John, Mark Smith and me to investigate and report back. John brought us together for the Target Taunton project over a series of excellent dinners in the Carlton Club in London, and we worked through the options. Mark examined it from the WSR viewpoint, I tested what might be possible with BR where I was working at Board Headquarters at Euston House, and John coordinated the work and constantly asked probing questions, usually starting: "Why can't we....?"

This work was carried out while BR was being privatised and the new structure and safety regime, including ROGS (the Regulation of Railways and Other Guided Systems) was being established. It quickly became apparent that the cost of becoming a licensed train operator on the Railtrack network was going to be high and the conditions onerous. Delays caused by a late-running WSR train could result in knock-on delays which could reach as far as Scotland with very expensive consequences under the compensation regime then being established. The additional requirements for running on the main line, such as train radio, black box recorders, train protection and warning system, and central door locking would all add to the start-up costs.

On the other hand, no such constraints would apply to main-line operators coming on to the WSR. The modern trains could quite happily cope with the heritage safety systems and semaphore signalling, and this led to the policy of encouraging main-line operators to come to us rather than the other way round. One constraint on running more trains was the rather cumbersome arrangements for transfer at Norton Fitzwarren. The closure of Taunton Cider in 1995 meant that it was no longer necessary to clear the wagons from the siding before any through train movement, but the sleeper barrier required a WSR signalman to be in attendance and the shunt signals leading to the main line to be flagged by Network Rail for any movement carrying passengers.

A scheme was developed with Network Rail and Somerset County Council to replace the shunt signal with a running line signal, and to replace the sleeper barrier with a WSR train token which was passed through a slot machine at Taunton station to release the signal at Norton Fitzwarren from the Down Relief line on to the WSR. Only one token is provided, and this arrangement prevents a main line train being routed on to the branch in error, or when a train is approaching from Bishops Lydeard. It was a struggle. The equipment in Exeter panel was then technically obsolete, so the number of changes that could be made was very limited. A late addition to the scheme was the sighting board behind the Up Main signal protecting the junction at Norton Fitzwarren, which was an extra cost but it was all sorted out in 2006, and we are indebted to Somerset County Council for meeting the £375,000 cost. The AWS ramp for Network Rail's up distant signal is on WSR track and maintained by WSR.

During my term as Company Chairman we dipped a toe in the water with the *Butlins Express* running during the summer peak between Minehead and Bristol on four days a week. Top-and-tail Class 31s added interest and we learned a lot from the trial. We were developing an improved offer for the following year, but in the event, the train operator Fragonset ceased trading and the idea foundered. Meanwhile, many charter trains and rolling stock movements, along with shuttles provided initially by BR then Virgin and latterly by Great Western have made use of the upgraded connection. The first HST visited the railway in 1993.

The principal user, though, has been Network Rail with regular deliveries of spent ballast for reprocessing at the Norton site. With David Holmes as Association Chairman, we negotiated a deal with Network Rail which worked for both sides, with a memorandum of agreement in 2006 which I signed for the railway. The result was stone for us and enough track to build the triangle, and a quick turn-round of the phenomenally expensive high-output ballast cleaner for them. So well has this arrangement worked that it is still in place, whereas originally we thought it would last only for 18 months with a possible extension to three years from 2006. It has also served to secure the continued maintenance of the main line connection itself.

We have never given up, and Paul Conibeare, Steve Williams and then Frank Courtney and I have been discussing the possibility of a seasonal shuttle with Great Western to bring visitors to the railway to extend our catchment area and to offer an alternative way of reaching Butlins for those with heavy luggage, buggies, bikes or wheelchairs. Initially plans were frustrated by the shortage of rolling stock and the desperate need to lengthen trains around Bristol and Exeter to accommodate the huge growth seen over the last decade. Now electrification has allowed enough units to be cascaded west and this shortage is not so acute.

An early use of the mainline link has been for freight, here seen bringing in rock for coastal defences. Allan Stanistreet

In April 2019 we were successful in securing a £60,000 grant from GWR's Customer & Communities Improvement Fund to support a scheduled Taunton - Bishop Lydeard link on peak summer days. This will improve social mobility in West Somerset, accessibility to top Somerset tourist attractions and provide some hard information on the potential for a regular link which has been the target now for over 40 years.

A Long Shift at Blue Anchor

Ian Coleby

I didn't really intend to volunteer for the West Somerset Railway. Although a Somerset lad living in Castle Cary, my original intention was to work for the Welshpool and Llanfair Railway in deepest Wales. At this distance and time, I can no longer remember why, though it was probably the narrow gauge atmosphere of that railway which attracted me.

I have been interested in railway history for many years and in 1982, I discovered the modernisation scheme that would sweep away all the mechanical signal boxes in the Westbury and Exeter areas. I immediately made it my mission to photograph as many of these as I could both inside and out. A lot of this involved tramping down muddy lanes or across wet fields to get the right view of the signal box, but I also managed to get some official visits with the area inspector who opened up many of the boxes which were switched out. I'm not sure entirely what it was, but these visits made me long to become a signalman and for the reasons which I've now forgotten the Welshpool line had become the target of my interest. I exchanged letters and arranged to start training in early 1983. At that point I dug out the map and realised quite how far Welshpool was from Somerset! Was there somewhere which was nearer where I could satisfy my signalling aspirations?

Then I discovered the West Somerset Railway. I was very quickly rostered as a trainee at Blue Anchor signal box and I met the duty signalman who was to be my trainer on a bright and sunny day in mid-June. A very pleasant and friendly man, I remember his first signalling lesson very well: "This is the staff ticket box. We always unlock this at the start of every day and leave it open in case we get the train staff stuck at the wrong end of the section." Those of you who are aware of signalling practices will understand the error in this statement, but I think enough water has passed under the bridge for me to be able to recount it now. The staff ticket box then was a simple green cashbox which you could unlock and leave open. Later it was replaced by a proper railway staff ticket box where the mechanism made this impossible. As it happens, he was an excellent teacher and I picked up the routine of the signal box quickly. It wasn't long before the signalling inspector was paying me a visit and I was, in short order, passed out to operate the signal box.

This was August 1983 and my training had been accomplished very quickly, evidence of my enthusiasm at the time. The railway was being run on a shoestring. At Blue Anchor, the museum hadn't been developed and the building was full of old bus seats waiting (so I understand) to be installed in a DMU, or maybe they had just been removed from one. There were no operational signal boxes at Minehead, Crowcombe or Bishops Lydeard and the entire railway was run day-to-day by the signalmen at Williton and Blue Anchor and a manager at Minehead. I would often work a shift in the signal box and see nobody all day apart from the train crew. In those days Williton was able to switch out and then I would have been the only signalman on duty.

I remember one particular autumn weekend. We often ran an early DMU that left Minehead at 09.00 or 09.30 on a Saturday morning, and as I travelled from Castle Cary this meant an early start from home to be there in time. This particular day was full of autumnal gales, and while Blue Anchor is a wonderful location on summer days, a strong easterly gale combined with a high tide can mean that spray from the waves lashes the signal box! On this particular evening, the *Quantock Belle* was running, and the lack of volunteers meant that it was normal for the daytime signalman to stay to see the train safely returned in the late evening. This meant very long shifts indeed and late-night journeys home. Throughout the day the wind had increased and by the afternoon I was finding it difficult to close the level crossing gates against the gale. I was advised that when the *Quantock Belle* ran through, the guard would help me close the gates so that I could go home.

I can't remember what time the train returned though it would have been somewhere around 10pm or so and by now the waves were breaking over the road along the seafront. The returning *Quantock Belle* reached Blue Anchor and sure enough the guard held the gates back against the stop in the road while I locked them securely. He returned to his train which then set off up the bank to Washford. The regulations being what they were then, I telephoned Williton to say the train was on its way, shut up shop, ran to my car and started my journey home. Without thinking what I was doing, I

The original Blue Anchor to Williton train staff. It is shown here with the non-standard 'cash box' ticket box that was later replaced with a proper GWR style one that provided much more security. Ian Coleby

Looks like a storm is brewing at Blue Anchor, the sort of weather that I experienced in this tale. Ian Coleby

came out of the yard at Blue Anchor, turned left and started off along the seafront. In those days I had a Mk2 Ford Escort which had little enough protection from floodwater underneath. I had only got halfway along the seafront when a huge wave broke over the sea wall and drowned the car which promptly stalled and stopped. No amount of effort would get the engine going again and I was marooned.

I decided that the only thing I could do was to return to the signal box and shelter there. It was a very wet walk back and I was pretty well soaked through when I eventually reached sanctuary. The good thing about Blue Anchor signal box, apart from the location, is that it has a very effective cast-iron stove, or at least it did then, and there was plenty of coal and scrap wood around the station. I soon had the fire stoked right up, I pulled the armchair as close to it as I could and sat there and steamed. The rain and spray were like grapeshot against the windows, which rattled loudly in their loose frames. Despite this, I must have fallen asleep and woke at 4am, stiff and cold, but it seemed that the wind had now abated. I relit the fire and wrapped the rug around me to keep warm and dozed again until it got light. By now things were calm and I had dried out. It was a pleasant walk back to my car where I had left it along the seafront. Luckily it had also dried out and I managed to restart it. It seemed like a long journey home which I reached more than 24 hours after I had set out the previous morning. Such were the early days of the West Somerset Railway.

Rather surprisingly, this experience didn't put me off. After completing a week at my day job, I was back there again for another shift the following Saturday. That was less eventful but was still another long shift and late night. The *Quantock Belle* must have arrived at Minehead at around 8pm and there had been some incident while preparing for its return to Lydeard. I received a phone call saying that the person who was due to operate Blue Anchor signal box the next day had managed to gash his hand badly and would not be available, could I therefore work the box again the following day? Unhesitatingly, I answered "Yes," and so 7.30 the following morning I was yet again on my away back from Castle Cary to the railway for the second shift of the weekend. None of this put me off, and I was soon also trained for Williton and later Crowcombe and Bishops Lydeard signal boxes.

I managed 30 years as a signalman on the railway before I had to stand down in 2013 due to pressures from my day job which meant I just wasn't able to commit to the shifts in advance. It is something that I very much miss, the opportunity to sweep away the years and to work as a signalman might have done 100 years ago is something I enjoyed very much.

Some signal boxes just won't stay still! Originally built for Hallen Marsh, it was diverted to become Nelson and Llancaiach East. After closure it was dismantled and moved to Dunster in 1934. The signal box had to be moved from Dunster as it obscured the crossing and as the WSR foresaw the need for a box at Minehead, it was yet again moved - this time by rail in an exercise masterminded by Hugh Perrett. Here it is seen being gently manoeuvred along the track to its new (and final?) resting place Nick Jones

Blue Anchor Railway Museum: A Dream Comes True

Ginny Barnfield

Seated at the kitchen table in an old Cornish farmhouse rented for our 1984 spring holiday Peter, very keen to set up and run our own museum, drew a plan with measurements and layout of his ideal building. Two months later, hearing that the WSR wanted a museum to celebrate GW150, we volunteered, met Managing Director Peter Rivett and General Manager Doug Hill and inspected the downside waiting room at Blue Anchor. The measurements conformed exactly with Peter's plan. Blue Anchor Railway Museum is the realisation of his dream. A logo was designed and work began.

Funding was minimal: £250 each from the WSR plc, the WSRA, and ourselves, so, on our first working day of the August Bank Holiday weekend a picnic table on the platform offered homemade marmalade, complete with the BARM label. When driver John Green arrived with the up DMU and called across, "Get the kettle on! This train has broken down!" a tradition was established: tea, free for a donation in the tapered old-style 17-gallon milk churn. Grateful passengers, waiting for a relief train, also bought all the marmalade. The adjoining caravan site ensured a steady market and once the damson crop from the trees overhanging the up platform ripened, jam was added to the range, then chutneys and suchlike. Our kitchen became a preserves factory.

That first weekend was spent clearing out the accumulated rubbish of many years including a Christmas tree, squatter's mattress, and a rotting crossing gate embedded in the brambles engulfing the up end of the building. This we sawed in half to the horror of Roger Woodham who had hoped to re-use it. Everything got added to a heap in the car park and we were assured it would be collected by train in a couple of weeks. It stayed there for at least two years!

Next job was to patch up and paint the walls, retaining the existing cream to lighten the interior. Williton stationmaster and Trust Secretary Chris van den Arend was most helpful in supplying light and dark stone paint in authentic GWR colours. For painting the ceiling and installing strip lights the DMU brought a scaffold tower down from Williton. Having never dealt with one before, needless to say we assembled it wrongly and the wheels would not move!

An alarm system was fitted, a stout padlock obtained and steel window shutters manufactured on site, courtesy of Mike Palmer from the S&D Museum at Washford. During three consecutive days in September, I added the skill of pop riveting to my c.v. A steep learning curve, this! The north-facing building was damp and very cold except in extreme heatwaves and the Company generously supplied two storage heaters, with power being brought from the up side of the station. For this, trenching was needed and one member of the gang was our son Ian, still a volunteer on the East Somerset Railway, from which we had recently defected. To bring the cable into the building the 2ft thick GWR brick walls were drilled by Ted Cubitt. It seemed to take a very long time before the bit made its appearance on the outside.

Thanks to Bill Love, Somerton signalman, we were fortunate to obtain cupboards which he had acquired from the box, and these too were painted light and dark stone, forming the corner on which were displayed the visitors' book, dip pen and bottle of ink, a donation box and another fundraiser, Peter's cardboard model of the building. It badly needed re-roofing, a slate roof was measured up and visitors were invited to sign a slate for 50p. I understand it is still giving good service 34 years on.

The initial shop was a trestle table, shelves in the window coming later. Preserves were prominent, we bought small items from the Company; John Green, the Watchet stationmaster, supplied boxes of second-hand goods, and one day Mike Walker, a regular holiday visitor from Coventry, appeared bearing a box of painted pebbles to sell. He kept us regularly supplied, carrying on the tradition begun by Mrs. Hembry, who came down from Williton on the first train, occupied the little back room of the station, raided the beach with her shopping trolley and painted and sold pebbles to fund the paint with which Ian Coleby decorated the signal box. Mike also made house signs at home to sell to work colleagues for BARM funds. We commissioned ceramic GWR fridge magnets based on a pannier tank and GWR-uniformed station cats. These were very popular and must by now be collectors' items! Tin trays featuring the GWR map also sold well.

We were occasionally given items to sell, memorably a biscuit tin of old Edmondson tickets for which Mark Bladwell gave us a very reasonable sum. Some months later he rang to say that a rare and valuable ticket amongst the collection had sold for a high price and he would send BARM a second cheque.

White-painted chipboard was used both for wall boards and showcase manufacture. Cut to size by our local supplier, corners carefully rounded by Peter, it filled the back of our long-suffering old estate car for many Sundays. The showcases were assembled with sheet plastic under the feet to protect from rising damp. Each was themed: Catering, Signalling, Uniforms, and Booking Office. By cutting a window in the back of Catering we were able to display a diorama Peter made, with a broad-gauge train illustrating Brunel's ideas. There was little emphasis on locomotives since live steam passed the door at regular intervals, visitors emptying out on to the platform to see the real thing, but a 1418 cabside plate was memorably delivered by former GWR driver David Rouse on an up train, bowling it along the ballast!

A quilt was made and raffled to support the museum and station at Blue Anchor. The quilters seen here are (L to R): Audrey Short, Mary Dowrick, Betty Knight, Ginny Barnfield. Peter Barnfield

Captions were set in 10pt Courier on my Smith Corona electric typewriter and pasted on to stout cardboard film boxes, readily available while Peter was still in full-time employment. Totally encased in damp-proof seal, they were stuck on to the chipboard and to my utter amazement some survive to this day, 34 years later.

Our friend Mary Gryspeerdt, Curator of the Somerset Rural Life Museum at Glastonbury, advised that captions should be aimed at a reading age of 12! But as we were also aiming to tell a story, these were in two stages, one giving basic information that was enough for many, the other giving a fuller narrative. Extensive use was made of photographs produced by Peter in our home darkroom. Professionally-cut polystyrene letters mounted opposite the door below Mark Bladwell's GWR clock spelt out GREAT WESTERN RAILWAY, in case of any doubt! Pre-computer interactivity was hands-on, with not a screen in sight. There was a superb model of a Cornish beam engine made and loaned by stationmaster Rob Lindley's father. It could be operated on payment of 5p and an Edmondson souvenir ticket could be stamped in the dating press, which only showed the year – everyone who has changed the date in a booking office will understand why: it was a fiddly, time-consuming job. The station wicket gate levers, on temporary loan from the signal box, could be pulled off, although not of course connected to anything.

We produced and photocopied booklets at home on the GWR and Bristol and Exeter Railways, a brief Station Guide, How the Signals Work, and a quiz sheet for children, while Peter designed colouring sheets on a railway theme and on canals. Once word got

Blue Anchor Railway Museum before and after. Above shows how it was found by Peter and Ginny Barnfield and below shows the carefully refurbished interior ready to receive railway visitors. 1985.
Peter Barnfield

around of the museum project, items for display came from all directions, some given, some on loan. Everything was properly registered and recorded in accordance with museum practice since BARM was being created under the auspices of the West Somerset Steam Railway Trust, of which Peter as Curator became a director.

The largest and heaviest item was the Waterlow's passenger ticket printing machine loaned by Edmondson ticket expert Michael Farr, who himself printed for the Talyllyn Railway Preservation Society.

Much work was done to the outside of the building, with a mock window painted on the blank Minehead-end wall and missing teeth of the valancing replaced. This was painted from a ladder on the track in February while I stood on the bottom of the ladder to steady it. We acquired various trucks and trolleys to add to the period atmosphere, with appropriate luggage, one item being a superb cabin trunk donated by a local resident. Genuine labels were procured, photocopied and applied, including a wonderful Cunard Line three-colour label proclaiming "NOT WANTED ON THE VOYAGE" by a couple travelling on the Queen Mary from New York to Memphis.

The museum at Blue Anchor was set up under the auspices of the West Somerset Steam Railway Trust. Here are seen the directors in January 1992. (L to R): BARM Curator Peter Barnfield, Trust Treasurer Chris Dowrick, Chairman Peter Thompson, Secretary Chris van den Arend. All sadly no longer with us. Ginny Barnfield

Large sack scales needed weights so I went begging to the Quantock Jam Factory in Bridgwater, and they were only too happy to donate some of their now-obsolete imperial ones. There were mail sacks, posters and enamel advertising signs on the walls to complete the air of authenticity.

Before we could open signs were needed to make people aware of the existence of the museum, and several coats of paint and layers of varnish ensured that they stood the test of time and weather. Putting them out was the second job on arrival. The first? Getting the kettle on! As Easter approached, we prepared to receive visitors, GWR uniforms were worn – Peter sporting earlier copperplate lettering on his buttons while mine bore the later roundel design. The kettle boiled, we opened the doors and waited. Visitors arrived in a small but steady stream. The museum existed!

The official opening came on July 14th, the ceremony being performed

Peter Barnfield is seen repairing the running in board at Blue Anchor. The previous plywood letters are replaced by proper GWR cast metal versions.
Ginny Barnfield

by our President, Major-General Errol Lonsdale, who succeeded Sir John Betjeman after he died the previous year. The Watchet Silver Band was in attendance, guests wore period costumes, refreshments were provided in the waiting room, there was dancing on the platform under the canopy, and a general air of celebration prevailed. We were grateful for offers of help to staff the museum as we were open every Sunday and Bank Holiday from Easter until September. From Bristol like ourselves came Mark Bladwell with his family, Michael Wyatt, and Stephen Park. Rose and Ted Wharnsby, from the Home Counties, spent three weeks in Minehead every year and did several days for us. On Bank Holidays, Richard and Betty Knight, from Wiveliscombe, took a holiday from running the *Quantock Belle* to help. Mary Dowrick, who lived just up the hill from the station, was a real stalwart, dealing with school groups, taking a turn on Sundays and opening the museum on summer Tuesday afternoons. She and Betty came up with the idea of a fundraising quilt which they and Audrey Short worked steadfastly to create. The squares were alternate patterned GWR brown and plain cream which took many hours to complete. I was allowed to add a token six stitches just to say I had contributed, but my main job was selling tickets on the trains. The stitchers, suitably attired in Victorian costume, posed for photographs before the draw, which raised a much needed £2,000 towards museum roof repairs.

The canopy was leaking and needed re-roofing with galvanised sheeting, the joints then being sealed with bitumastic paint. This job was carried out by the two Peters (Thompson and Barnfield). Friends of Blue Anchor Railway Museum was another fund-raising idea with a £5 annual subscription conferring no privileges other than the satisfaction of supporting something in which you believed, and a £100 life membership. The WSR was always good at celebrations but I was flabbergasted when, during the 1992 Steam Trust AGM held in the BA station waiting room, the up *Quantock Belle* made an unscheduled stop, chairman Peter Thompson adjourned the meeting and in burst a contingent bearing a huge bouquet, cake and a card signed by almost everyone on the railway. Now who had let on that it was my 50th birthday?

We carried on running BARM for another two years and were very grateful to Pete Treharne, already the stationmaster, who then took over as Curator. The WSR played a very important part in our lives for 10 years and I would like this article to stand as a tribute to Peter who died in August 2018, with thanks to our son Ian for his proofreading and editorial advice.

The Tale of a Blue Anchor Stationmaster

Rob Lindley

In 1984 I was appointed stationmaster at Blue Anchor, where I was a signalman and Andrea, my fiancée, was training in the box. She was not the only woman but certainly the youngest to achieve the role and on her 18th birthday she signalled for the Wednesday evening Sunset Special, which came in bearing a headboard "Happy 18th Andrea!"

Peter and Ginny Barnfield had already started to create the museum on the down platform and for the next eight or nine years we worked as a team, aiming to restore the station as far as we could to give an authentic GWR atmosphere. Ian Coleby had taken on the task of painting the signal box and in due course made a new nameboard, cutting out the letters laboriously with an old bandsaw. He went on to make the "Ladies" and "Gents" hanging signs, the destination board with the fingers which slotted in and out and also refurbished the catch points sign board, touches which made an immediate difference to the appearance of the station.

Meanwhile we started on the main building. Once we had cleared out the old furniture from the booking office, we painted it cream for lightness, while the stove got a coat of black and looked as good as new. Graham Stagg supplied a period calendar in polished wood, some old box files were stacked on the shelves, and it began to look like a railway booking office rather than a tip.

The signalman's job included selling tickets, a notice on the closed booking office window telling would-be passengers to apply at the box. He also had to advance the date on the Edmondson dating press with its inky ribbon, a messy job for beginners. On busy days such as Bank Holidays, Ron and Audrey Short would share booking clerk duties and our aim was to have the office manned as much as possible. New glass was cut for the arched sliding window opening and we acquired the ticket barrier from Somerton, plus the lovely brass GWR escutcheon plates. Fitting the barrier was a difficult job, since when I drilled through the floorboards to accommodate the very long metal uprights, I hit a flagstone floor just six inches below and went through at least half a dozen hacksaw blades to get the uprights short enough!

Ian Coleby made "In" and "Out" signs from wood, later replaced by the authentic cast ones, and a leaflet rack from varnished plywood. I found some lightshades and put up wall lights which, although not actually GWR, added to the period feel of the waiting room, while Ginny and Peter reupholstered the padded bench, from which the horsehair was escaping. Chris van den Arend had taken scrapings from the GWR green paint used in the waiting room and found that Woolworths sold an exact match, which was duly applied. The transformation was amazing.

Then it was time to tackle the outside and here we got help from various quarters. Bob and Joy Shortman had a caravan just over the railings and asked whether they could cut back the laurels obscuring their view of the trains. I was really pleased because Joy then offered to take on the up side bank. This was very bare, and it was transformed, firstly with variegated periwinkle from Peter and Ginny's Bristol garden which spread rapidly, as can still be seen, and later Joy added more and more plants which she brought on at home. For many years, when they were not staying in the caravan, Joy came down every Sunday from Saltford, near Bath, to maintain what had become "her" garden. Bob went on to refurbish the Ladies, including the roof and a replacement finial.

The booking office window facing the platform was rotting and this time help came from a holidaymaker on the caravan site who did not like just sitting on the beach with

Rob designed and built the replacement station shelter on the up platform at Blue Anchor. Here it is seen in a semi-complete state. Peter Barnfield.

The results of the hard work of Rob Lindley and the many other volunteers who have contributed to restoring the station at Blue Anchor are seen here in June 2006. Ian Coleby

his family. Dave Taylor, a professional carpenter from Radstock, repaired it by replacing the bottom third of the frame and the sill, which was mainly filler, making such a good job that you can hardly see the join today. The scruffy plywood letters on the up side running-in board were a source of annoyance, so when Malcolm Broadhurst, bringing a visiting party from the Severn Valley Railway, offered to provide cast iron replacements in return for borrowing some of the cast letters that we did have to cast the ones they needed, I jumped at the chance. Yet one more improvement completed.

To alert visitors that the station was now very much open, Peter lettered up a large sign and I gave it about six coats of varnish to protect it from the seaside weather before we put it up on the back of the signal box facing the road. It still looks as good today. As with the museum, the station needed to raise funds and I remember Peter turning up one day with a huge bag of empty 35mm film containers which we turned into cash by filling with smokebox ash and sticking on labels saying it was genuine ash from our GWR loco, Pannier 6412. We sold them for 50p and made a lot of money. Another fund-raiser was coasters of North Wales slate depicting various locos. The annual BA mug, eagerly collected, came much later and was initiated by Pete Treharne.

Money also came from our supporters. The station roof had long needed attention but it simply wasn't in the Company's budget so Stuart Bunyan, who lived nearby, came to the rescue with what turned out, to my embarrassment, to be a very long-term loan. Stuart will always be remembered for his "scorched earth" policy on the bank beyond the museum. We were all horrified at the result initially, but of course he knew what he was doing, as is proved by the way it looks today. Another task on that bank was getting rid of a wasps' nest by smoking them out one evening and then digging out the nest.

The biggest building project was a shelter between the booking office and the ladies which my dad helped design and I built. Peter helped me lift the roof timbers in place, and all the materials were delivered to the station yard, then moved manually to the platform with a trolley on the line which we piled with breeze blocks and then pushed over the level crossing to the platform. It took quite a few trips and I think we used about 800 blocks. I remember re-plumbing the station water supply and then received a complaint from the owners of one property that they had no water. It turned out that they had been having free water from the railway for years! The signal box door from Wellington replaced BA's rotting original, and we also stripped out most of the furniture from there at the same time. Another trip saw the furniture taken from Taunton West Signal Box, along with the linoleum that now graces the floor at BA. Some of the excess furniture ended up in Crowcombe signal box. We had the contents of Silk Mill signal box as well. During one of the Sunset Specials on a Wednesday evening, a woman asked me if she could look in the locking room under the signal box. It turned out that during the war her job had been to maintain all the equipment batteries in the signal boxes on the Minehead Branch.

I also organised a trip to Bristol, where Robin White had arranged for us to acquire some GWR cast iron lampposts which had come from a museum at the old Bleadon and Uphill station then ended up being in store at the original Temple Meads Station. I drove the flatbed lorry to collect them and held up three lanes of traffic while trying to reverse through the narrow gateway at the side because I had only about six inches of clearance! On an overnight working party at Starcross station we dug out more cast iron lampposts during an engineers' possession and there were no trains running. The last one was being slid down the main line rails to the barrow crossing while the BR manager was frantically trying to get us to sign back the possession for the first trains. Swapping with ones at Bishops Lydeard and Williton ensured that each station had the same style, BA getting earlier posts to match the ones from Bristol. Some of the original steel Tilly lamp posts were succumbing to salt-water corrosion while much taller concrete posts, cast at the GWR's own Taunton concrete works, were removed when the top of one fell on to the platform. Salt air and metal reinforcing rods don't mix! Installing the cast iron posts required much trenching work before the cabling could be laid, and I am grateful to all those who helped. We had to break up bedrock on the down platform because it was only about three inches below the surface but the up side was not as bad and a lot easier. Lamp tops were fabricated by an engineer from Honiton who was so enthusiastic about the project that he gave us a very generous discount and also designed the bracket which holds the lamp under the museum canopy. The Blue Anchor lettering was sourced in Bristol and applied by Peter and Ginny.

The signal lamps were all still paraffin and on windy days we had to struggle to light them at the foot of the exposed up home signal next to the beach, then climb the post with the lamp in a bucket, quickly transferring it from bucket into lamp holder before it blew out. My record? 5 attempts!

I finally resigned as stationmaster in December 1991 and handed over to Pete Treharne before my first offspring was born two months later.

I Didn't Mean to be Managing Director

Mark Smith

Mark was Managing Director of the West Somerset Railway from 1988-2006 and has continued volunteering in various capacities since. He is now a director of the PLC.

I came to Somerset in January 1984 to take charge of a Taunton primary school and wanted to get involved in practical railway work again. I approached the WSR and, after two months passed without any reply to my letters asking about opportunities to train as a passenger guard or a letter written from my school asking for details of party rates for educational visits, I decided to take action.

A door on the platform at Minehead station headed "General Manager" seemed a good place to start seeking answers. A voice called me to enter, and as the door swung open, I knew immediately why my enquiries had not been answered. A solitary man sat at a large desk deeply laden with paperwork. More files were stacked on the floor and various shelves. Much rummaging among the papers revealed the school's enquiry, and that was sorted. My volunteer application with a rather scrappy piece of quarto paper attached, issued 20 years earlier and certifying me competent to work as a guard on "unfitted passenger and goods trains only" by the Welshpool and Llanfair Light Railway was never found.

It quickly became obvious that while plenty of people wanted to be part of the operating railway, no one wanted to help the general manager with the essential paperwork, so I volunteered. I quickly established a routine during school holidays of travelling to and from Minehead by train and helping to lower the stacks of paperwork in the office. At about the same time I joined the WSRA. I was told in no uncertain terms that as I could clearly afford to pay fares, I was not to use the free allocation that came with WSRA membership. The Company needed the money. Not a lot has changed!

It was at a WSRA meeting that Richard Knight introduced himself and wondered if I had ever thought of working for the *Quantock Belle*? I learned how to lay a table and serve silver service but found bar stewarding more fun. One Saturday evening Barry Malcolm told me he was on his own in the kitchen and I had to help. Cooking dinner for 73 in a tiny kitchen is a very different experience. We had a blowback with the oven, I lost the hair on my right arm and Barry lost an eyebrow. Dinner was apparently edible, but I wasn't asked to cook again. The role of Association secretary beckoned and that meant a place on the Company board.

By 1988, negotiations for a new lease of the line with Somerset County Council were advancing. I found myself as the local director making up numbers as the details were discussed. This new lease would give us security for 99 years. It committed the company to maintain the line and operate 400 trains each year. The Company board wrestled with the responsibilities, which included the need to raise £210,000 to purchase. Peter Rivett, then Managing Director, resigned as he could not support the proposals, and to fill a vacuum I offered to become volunteer Managing Director. Changes in education legislation were giving head teachers responsibility to manage schools as businesses, and the local education authority thought managing a heritage railway would be good practical experience. I was given two days' leave per term.

In 1988 two pieces of good news came to us: the National Railway Museum would hire us No. 92220 *Evening Star* for a complete year; and Councillor Humphrey Temperley, leader of Somerset County Council, discovered a regulation which allowed a local authority to keep part of the proceeds of any asset sale. He arranged for the County

The visit of 9F 92220 Evening Star *in 1988 was a major turning point in the fortunes of the railway. Here the locomotive is seen at Minehead hosting local MPs Tom King (now Lord King) and David Nicholson. Also seen is railway enthusiast MP, the late Robert Adley.* Courtesy of Lord King

The significance of the visit by Evening Star *in 1988 cannot be underestimated. No trains in this picture, but it shows just a small part of the crowd that gathered to see the locomotive. Here Crowcombe Heathfield station sees more people than probably at any time in its past.* Walter Harris

Council to purchase shares in the Company to the value of £50,000. We now had to find only £160,000 and we had Swindon's finest as an attraction.

The Railway Inspectorate were rather concerned that we were taking on such a heavy and prestigious locomotive. Major Olver of HMRI gave official sanction but privately bet me £5 that the locomotive would derail. When *Evening Star* eventually left the railway, he rather glumly handed me my winnings. The much bigger prize, however, was the number of passengers attracted by the loco.

Transfer of the lease was completed on 1st May 1989. We could now plan for the long term. Priorities were a shed so that carriage maintenance would no longer be weather dependent, and to discover what the real capacity of the WSR's infrastructure might be. In the meantime, the Managing Director responsibilities were proving too onerous. I gave notice and supported the plan to recruit a full-time Managing Director. Job description and contract of employment were written and the post advertised widely. Response was poor, I guess the early history of the WSR with its financial crises deterred the cautious and sensible from applying. I recall explaining to my wife Pat that I felt I might be able to do the job as well as any of the applicants we had seen so far, and to my great surprise her response was: "Go for it, it's where your heart is." In view of my involvement, the whole board decided to participate in the interview process, and I endured a couple of hours' grilling in the upper room of the Egremont Arms at Williton before accepting the post.

The Great Western Railway, and later British Railways, had managed all traffic on the branch using relatively light axle-loading locomotives and we could find no evidence anywhere that any surveying had ever been undertaken to confirm a higher axle load other than for the section Norton Fitzwarren to Bishops Lydeard. We engaged an independent consulting civil engineer and set off very early one morning in the DMU from Bishops Lydeard with ropes and ladders to assist access. Two long days later after

52 bridge and culvert inspections we arrived at Minehead with a lot of optimism because axle loading of 23.5 tons was thought possible with only a relatively short list of minor works required. We then set about measuring our rail depth and sleeper condition. The independent track engineer's report was more cautious. His recommendation was for a slightly lower tonnage with occasional passage of 23.5 tons possible.

While awaiting confirmation from HMRI of their acceptance of the two reports, we were lucky to have BR standard No. 71000 *Duke of Gloucester* stabled at Bishops Lydeard following a main-line charter. A rather helpful provision in the Railway Safety Case Regulations allowed for the testing of new equipment and so it was decided to run "The Impossible Dream," a charter from Bishops Lydeard to Minehead hauled by the Duke. The train was arranged and sold within 10 days without the benefit of the internet and just by word of mouth. We shared the profit with the locomotive's owners.

At last the new direction as to "Speed and Weights" issued by the Secretary of State for Transport arrived, giving the freedom to invite any surviving ex-BR locomotive to work the branch. We often boast about being Britain's longest heritage railway and operating more stations in Somerset than BR – useful in our marketing but all those miles and stations need maintenance! We needed more income to pay our way. Our annual timetable brochure was the principal marketing tool and we argued long and hard about where to place our advertisements, newspaper or magazine. Increasing business in the shoulders of the season was another pre-occupation. The first Thomas events were offered in an October half-term weekend.

No matter how much income increased, our increasingly elderly railway soaked up cash. The annual budget must support the operation of the line and its supporting administration costs. Most of the income is earned in the summer while most of the expenditure on repairs and maintenance is in the winter. Cash flow control is all-important and very frustrating. There are always heart-rending appeals for some wonderful heritage or historic development while the reality is sleepers, ballast, concrete, roof slates, fencing, and the boring essentials like insurance, electricity, photocopier hire. As Managing Director, I enjoyed the support and control of a board of directors who had real experience of the difficult very early days when the company was in serious financial difficulties. Rigorous financial control meant saying No to many pet projects.

However, all our researches indicated that with no increase in the number of tourist beds in West Somerset in prospect, growth could only be achieved from the southern end of the line. It is worth remembering that in GWR and BR days, Bishops Lydeard was very lightly used. There were no locomotive facilities of any kind, and Great Western locomotives must go over a pit for essential oiling and inspections. An inspection pit was dug and protected with chain-link fence. The BL compound enabled locomotives to be stabled overnight. For the first time in preservation, steam services originated from the station. Traffic grew.

Who can remember how we coped? Train movements were managed by ground frames at each end of the passing loop, a single telephone line was shared by operations, commercial traffic, and of course the Association. The platform extension, shop, café, booking office and signal box to control train movements that we all take for granted were but dreams in those early days.

The Swindon Shed

Mike Johns

Mike Johns worked as a British Railway engineer from 1953 to 1992. He was closely involved in the Braunton *project at Williton from 1996 and still maintains a close interest in work at the Swindon Shed.*

During the late 1980s, Tarmac Ltd was the lead contractor in the clearance and development of what had been the Great Western Railway's Swindon Works. Although certain parts were to be retained because they had a particular historical significance or there were new uses for them, a building known as the Machine Store was in the way and needed to be removed.

This was Grade 2 listed and therefore could not be demolished but could be re-located. The reason for listing this two-bay building was the design of the roof over each bay which had a curved profile supported on wrought iron truss girders clad in corrugated iron sheets. This was the first roof of its type in the world, hence the listing which required Tarmac to move it to a new site because it wasn't wanted in the planned Swindon re-development.

Tarmac offered to re-locate it at their expense to any heritage railway that had a potential use for it and adequate space in which to place it. At the time, the West Somerset Railway was sorely in need of covered accommodation for the maintenance of its increasing fleet of locomotives and coaching stock and, importantly, had a vacant open space inside the railway fence to the north-west of Williton station. However, the building itself measured some 200 ft long by 100 ft wide and would not have fitted between the Down running line and the fence in its existing form.

The outcome was an agreement that Tarmac would re-locate one bay only of the building to Williton at their cost, which preserved the associated curved roof design and hence its Grade 2 listing. They cleared and levelled the present site, installed appropriate foundations, re-erected the building and left it weathertight. Around 1990 the WSR assumed responsibility for its future management, repairs and maintenance, and for fitting it out to suit its new purposes whatever they were to be.

At the time, the WSR locomotive department had the use of the old Minehead goods shed, which could accommodate one locomotive only for maintenance work, leaving any restoration projects to be carried out largely in the open air. The WSR Association owned small prairie tank No. 4561, which was operating in service, and pannier tank No. 6412, which was being overhauled by a WSRA volunteer team in the open at Minehead and was at the mercy of the weather. It was therefore agreed that the WSRA overhaul projects should re-locate to Williton to become the first occupants of the Swindon Shed, alongside the West Somerset Steam Railway Trust who had a similar need for covered accommodation to progress their main project of the time, the restoration of the ex-GWR Dean sleeping car of 1897.

Two tracks were laid through the length of the building served from Williton goods yard part of which, including the goods shed, were leased to the Diesel and Electric Preservation Group. In the building, a plain track was laid next to the running lines (No. 1 road) but the second track (No. 2 road) included a 60-ft inspection pit. The rest of the floor area was made up with ash. In 1993, No. 6412 was moved from Minehead and took up residence on the inspection pit for its overhaul to be completed, and the WSSRT team settled in at the opposite end of the shed to continue restoring the sleeping car. The WSRA team also brought one ex-GWR Toplight coach and two ex-Southern Railway PMV vans from Minehead. These were parked on the line connecting the pit road to the South Yard tracks occupied by the DEPG. The coach provided basic messing and office facilities plus some compartments for sleeping accommodation for volunteers wishing to stay on site. The PMVs were used to store tools and materials.

The Swindon (Tarmac) Shed is seen here in 2004 after the North Yard developed. Allan Stanistreet

The interior of the Swindon Shed is equipped with heavy engineering plant and able to tackle some of the largest jobs on the railway. Mike Johns

Subsequently the coach was displaced and Portacabins provided in its place.

These arrangements were formally recognised by the WSR in January 1994. By 1997 the WSRA had concreted part of its plain floor area to allow work benches and some machine tools to be installed and used safely. Rebuilt West Country Class locomotive *Braunton* had been purchased by the Association and was due to be moved from Bishops Lydeard to Williton where workshop space would be required once its restoration started in earnest. An agreement was made for the joint provision of drainage and electrical supplies, with separate meters for the WSRA, WSSRT and DEPG, to level and compact the outside ground level at the station end and to maintain access via the station yard and DEPG areas to both ends of the shed for emergency vehicles. Much of the funding was provided by the WSRA and WSR.

By this time most of the donated machine tools had been installed and workbenches provided over the length of the WSRA shed area with some of the heavier tools only used intermittently alongside No.1 road, along with more benches. Most of the machine tools were donated to the railway and placed in WSRA care. The whole idea was to provide a heavy repair and overhaul facility for locomotives and coaches, to complement the facilities at Minehead.

Other minor works were undertaken with funding from what was known as the Iffy Rivet Company, which was originally formed at Minehead as the volunteers' tea club. In time this generated surpluses to which were added income from donations on open days and by private visitors. The most notable contribution was for the replacement of the early cabins by ones for office and mess room accommodation plus a meeting room and toilet unit, which are still in use.

The number of volunteers working allowed the workshop to be open seven days a week plus Wednesday and Saturday evenings. The purchase of *Braunton* virtually exhausted the WSRA locomotive fund, and a dedicated restoration fund was set up with an appeal for donations to allow work to begin. One of those who expressed an interest in the project was Jeremy Hosking, who offered to lodge a significant sum in a bank account as working capital in return for title to the locomotive, which became the first of his fleet. A formal agreement was made in December 1996 between Mr. Hosking, the WSR and the WSRA, and provided for the WSRA to undertake the work and manage the locomotive for 10 years in operation after completion.

Importantly, it was recognised that a full-time project engineer would be needed at Williton as work on the locomotive was to start in March 1997 and Gareth Winter was appointed to this role. At the time, No. 6412 hadn't been completed and didn't go back into traffic until 1998. Initially the volunteer team were able to cope with the work but had limited skills when it came to efficient machine-tool production of new or re-furbished parts. A machinist was appointed to the permanent staff on the understanding that when not working on *Braunton* he was to be used on contract work to cover his cost. This led to today's mixed workforce and the work that has been done for outside customers when there were no WSR or WSRA requirements.

Over the years there have been installed an overhead 5-ton crane, which runs through the south end of the workshop, and four 25-ton lifting jacks from BR's Bath Road depot in Bristol. These allow any rail vehicle up to 100 tons to be lifted for attention to wheelsets, bogies, and so on. They can be used inside the workshop or on a concreted area of track outside.

The Swindon Shed is still the centre of activities but can't provide for all needs. Today it is a very valuable asset on the railway, and we should be grateful to Tarmac for their generous donation.

The Steam Trust's restoration of the 1897 first class sleeping car was largely undertaken inside the Swindon Shed. Here it is seen in a 'state of undress' while the exterior wooden panelling is renewed.
WSSRT Collection

Surveying the Heritage Coaches

Claire Sheppy, MRICS

As far as I know, my role on the West Somerset Railway is unique. My association with the Minehead branch began long before it became known once more as the West Somerset Railway. In 1953, my family - mum, dad, and with me as the oldest of six – paid our very first visit to Dunster Beach, staying in a chalet. It wasn't long before we, with a very railway-minded father, soon started to explore the line. In October 1958 my mum succeeded in buying one of the chalets at Dunster Beach and from then on, Dunster station became a regular port of call for the whole family, a situation which remained until the death of my parents, mum in 2013 and dad in 2014.

Our first active involvement with the West Somerset Railway, as it became known, was when my mum and I both bought shares in the infant heritage railway in 1976. Not long after, dad became a member of the West Somerset Railway Association, and in due course I too joined. I was living in Surrey but visits to Dunster to my mum's chalet continued nonetheless. In 1978, I had the opportunity to relocate to Somerset, moving to Bridgwater, but with a young son and later a second son on the way, visits to the WSR were somewhat sporadic.

Thus it wasn't until 1995 that I became more fully involved, mostly at that time by supporting the Spring and Autumn Galas. All this time mum and dad were still regular visitors to Dunster, as mum retained ownership of her chalet until 1986, but even after that they would regularly visit Dunster and Dunster station, staying locally in their motor-caravan. They became quite well acquainted with Chris and Colin Dowrick, joint stationmasters there at the time and it was as a result of this friendship that my parents donated the Severn, Avon and Wye poster to Dunster station, which can be seen in the station waiting room today.

On one visit to Dunster station I was with my parents when Chris Dowrick asked me where I lived. I said in North Petherton and he asked me if I would be interested in helping with the Heritage Carriages Project which aimed to restore enough coaches to

Measuring one of the heritage coaches prior to Claire's production of detailed drawings. Claire is here being assisted by Jennie Dunse. Mike Dunse

produce an authentic train of GWR coaches and which was still very much in its infancy. "It is very easy," he said. "It's just a matter of wielding a paintbrush," or words to that effect. As a result of that conversation I got it in touch with the West Somerset Steam Railway Trust, and in due course received a letter from Chris van den Arend saying that I would be welcome to join them. On Wednesday 25th April 2007 I travelled to Williton to find a group of about eight people pulling TK 3639 apart as fast as possible.

While my main passion is contemporary music and an interest in railways is one of indoctrination, my qualification is as a Chartered Surveyor, a career that I had always regretted choosing apart from one thing: the aspect I enjoyed most as a result of my training was constructional drawing. At this point I began to wonder how they were going to know how to put the coach back together again, so I asked if they had any drawings. Just one was produced, and I distinctly remember saying to them: "Well, you have got the coach, I will measure it for you and produce a set of drawings." From that moment on, I have done very little in the way of actual work on the coaches themselves, but bit by bit and visit by visit I measured all the key components of the superstructure and produced working drawings of a Toplight coach based on my findings on TK 3639.

Once I had more or less completed that set of drawings, and with nobody telling me to stop doing it, I then went on to produce drawings for some of the other coaches, a job which is still far from complete. And occasionally in the early days I would produce a display of these drawings, one being at the re-launch of No. 34046 *Braunton* on 24th September 2008. On this occasion, I had displays of the drawings at both Bishops Lydeard and Williton, where I was able to show a number of people around the coaches that we then had awaiting restoration. Among those present was Danny Hopkins, then editor of *Steam Railway*.

As time went on, one coach became my firm favourite: BTK 7538, the Toplight Tri-composite brake coach which came from the Chinnor and Princes Risborough railway and is the second oldest coach on the West Somerset Railway after the sleeping car, having been built in 1907. As it was feasible to work on only one coach at a time – well, perhaps two at the most – I suggested that all of the coaches awaiting restoration should have their own steward. Of course, the coach I wanted to steward was BTK 7538. I pay fairly regular visits to Williton to keep a watching brief on it and to renew the polythene which prevents the worst of the rain getting in but allows a good airflow through it.

While carrying out stewarding duties at Williton at one of the Galas, I was offered on behalf of the Trust some original GWR blueprints from the LMS Society. Two of these have proved invaluable in helping me to prepare a set of drawings of the bow-ended corridor coach TK 4546, for which the drawing is still ongoing.

Initially I had only an A3-size drawing board and it was clear this was going to be inadequate, so I put out an appeal via the WSR.org website for a rather larger drawing board. I was somewhat flabbergasted when I was offered a total of six. I can clearly remember my good friend Ian Coleby remarking at the time that I had been offered enough to start a drawing office! Of these I selected two, one of which I intended to

Claire is seen here drawing up her survey of the statutory services in the Williton Sherrings yard for Gareth Winter. Gareth Winter

have at Williton and the other at home. One had been offered by IBM and the other by Fairey Hydraulics Ltd. This I installed in my conservatory, but in 2010 I decided that I was going to design and then build an extension on to my house which was going to mean the conservatory would be demolished. I planned a drawing studio/library in my garden, complete with heating and lighting, which would have to double as a store for the various items needing to be rehoused during the extension work. As I did so much of the design and construction work myself, my involvement with the WSR became rather less for a time, but now all that work is finished and after coping with the deaths of my parents, I have once more turned my attention to producing a set of drawings for TK 4546. After that I am hoping to produce an accurate set of drawings for the coach which I steward.

Two of my lesser-known involvements for the West Somerset Railway have involved producing a complete survey of the position of the statutory services such as electricity and water mains and drainage runs in the Sherrings yard at Williton, and the setting-out of the levels of the formwork for the Williton footbridge, then checking levels after the concrete pour.

I have also become instrumental in obtaining publicity for both the WSR and the WSSRT as a result of my ongoing involvement with BBC Radio Somerset, on which I regularly participate. In the summer of 2018 I had the pleasure of escorting both Simon Parkin and Charlie Taylor to Williton and back by train for Simon to carry out some interviews including one with me talking about the heritage coach restoration project. These were broadcast over six days during the second week of August 2018.

Track alterations in hand at Bishops Lydeard. Early 1985. Tony Yardley

Maintaining the Tradition: Edmondson Ticket Printing

Ginny Barnfield

The largest and heaviest item in the Blue Anchor Railway Museum (BARM) collection is undoubtedly the Waterlow passenger ticket printing machine. The first loaned by Edmondson ticket expert Michael Farr, who himself printed for the Talyllyn Railway Preservation Society unfortunately lacked the all-important attachment for printing serial numbers needed for accounting ticket sales. But through his contacts at the BR ticket print shop at Crewe he was given the vital part although it was clear we would need the expertise of a printing engineer to fit it. Michael arranged for the machine to be lent to Amberley Museum in West Sussex, who have returned it to working order and now use it for printing tickets for their own narrow gauge railway.

A pair of the special Edmondson tickets printed at Dunster.

Michael then bid and paid for a replacement machine from Crewe, which he generously donated to BARM and it was delivered to the museum by train from Minehead, Paul Conibeare operating the GWR hand crane with great skill. It just wasn't possible to lower the jib sufficiently to get the machine right into the museum, to which a very slight nick in the architrave bears evidence, so it was manoeuvred in by hand on steel rollers.

Unlike some heritage railways, the WSR has continued to use Edmondson tickets, which are such a traditional part of the passenger experience. The ticket press now in use was manufactured in 1944 by Waterlow and Sons as machine number 717, and is one of about 36 similar ones that survived although only six actually worked until the end of British Rail's ticket printing at Crewe, which finished on Christmas Eve 1987.

The Company was advised to bid for this particular machine after a review of the remaining machines by Michael, who also gave much assistance during the purchase negotiations with British Rail. Spotting a counting and binding machine, Commercial

Manager Hein Burger promptly bid for and acquired this as well, an invaluable addition to the operation, saving many hours of laboriously bundling by hand.

After much preparatory work in the former parcels office at Dunster which, among other things, involved the installation of 3-phase electricity and a large concrete plinth complete with large retaining bolts – for which the floor had to be reinforced – the machine was installed. After only an hour's tuition by Mick Norman, who was formerly in charge of the Crewe ticket printing shop, the first tickets were printed on 3rd January 1990, these being a batch of 1,000 for Blue Anchor.

For many years printing was carried out by Chris Dowrick with some help later from twin brother Colin when he retired to West Somerset. As parts on the operational machine at Dunster wore out, our machine was cannibalised, with Michael's full permission, to keep it going. The ticket printing plates are still obtained from the same firm, now based in Liverpool, the only major technological change having occurred eight years ago when images began to be sent by email rather than being printed on paper and posted to the block maker who then had to scan the images.

Over the years the machine has worn out several sets of rubber rollers and has had to have a replacement 3-phase electric motor. For at least the past 20 years most of the blank Edmondson card has been obtained from a firm based in Holland. During November 2018 the total printed since 1990 passed 15 million, providing a profit-making service to many heritage railways and other concerns.

Grateful thanks to Michael Farr and Graham Robinson for supplying the information without which this article could not have been written.

The Waterlow ticket printing machine on display in Blue Anchor Railway Museum. Peter Barnfield

"We Never See You Doing Any Work" Tales of a Lineside Volunteer

Paul Ritchie

On retirement from an office-based job in 1998, I joined the Cutting Back Gang (CBG) under the expert leadership of the West Somerset Railway Association's Special Projects Officer Ian Jonas, affectionately known as 'Foggy'. Manual work outside in the railway environment where the impact of our labours is so beneficial has a strong appeal for me. In the early years of the WSR, trees and bushes encroached on the trackside, so it was then a huge task to improve the environment.

The CBG meets on a Saturday and Sunday each month. The qualified chain-saw operators deal with the "big stuff" - felling and log cutting - while the rest of the gang use bow-saws to cut the smaller branches which are then dragged to a bonfire site. Trees which are unhealthy or impede visibility are selected for felling, leaving healthier specimens to grow more strongly. This thinning process reduces the shade and allows flowers to flourish under the remaining canopy. It also improves the view for passengers and locomotive crews and visibility at footpath crossings.

Logs are stacked lineside to await collection. This might seem straightforward, but it isn't when they have tree trunk dimensions at the bottom of a big embankment! To help this task our intrepid leader devised a "wooden railway" in several sections bolted together on which a small horizontal platform on wheels is rope-hauled up to track level with a payload of two or three heavy logs. This contraption became known as "Foggy's Funicular" and it replaced the rather precarious chain gang method of passing each log to the top of an embankment.

When trains are running, special attention is given to ensuring the line is clear in good time before a train is due. When one approaches, all work stops. The warning whistle or horn is acknowledged by all with a raised hand, signifying that we're in a safe position and the line is clear. It is then customary for us to do our PR bit by waving to the crew and passengers. This can lead to accusations that "we never see you doing any work." That is, before the bonfire smoke infiltrates the open windows on the train!

Naturally, passing trains provide the opportunity for a break from our labours (and to take photographs). On one occasion at Nethercott, we could hear the engine of a down train struggling for adhesion on damp rails until the Manor finally stalled on the curve in the cutting. We were pleased to assist sanding the line, the Manor found her feet, and the train was soon on its way to Crowcombe. All part of the service.

All volunteers are required to attend and pass the Personal Track Safety Course and sign a declaration on arrival at the work site, following a briefing by the Person in Charge (PIC). On site, safety is our watchword, individually and collectively. There are many hazards to be aware of, apart from falling trees. Wooden sleepers can be very slippery as can the steep sides of embankments and cuttings, brambles lurking on the ground act as a trip hazard, and inevitably there are rabbit and badger holes. The down side cutting at East Combe is so steep that a rope was tied to a tree trunk to aid the climb from track level. It was great fun abseiling down.

Another challenging location was the steep bank behind the platform at Stogumber where the natural fall of trees was down to the lane below, with telephone wires in between. David Holmes was then our intrepid leader and to his credit this very tricky job, which was not for the faint-hearted, was completed with minimal damage to the infrastructure. Sometimes our clearance work reveals something of particular interest such as a culvert at Churchlands and a platelayers hut on Washford Bank.

Logs stacked by the line are collected from time to time by a log train gang. This may be steam or diesel hauled, usually consists of several bogie bolster-type wagons and two or three brake vans to travel in. The job isn't completed until the logs are offloaded, and on at least one occasion the logs were thrown directly into farmers' trailers, quantities and value assessed, and monies agreed, all of which benefits the railway. Similar arrangements apply for the occasional Tidy Train when all sorts of material is collected from the lineside, such as chairs from replaced sleepers for re-use or scrap, rubbish dumped "over the fence" and so on.

Every cutting-back day involves a bonfire or two, managed by Mike "the fireman" Jenkins. Positioning is significant relative to the work site (as close as possible), wind direction (avoid smoke where the gang are working), and nearby trees, bushes, wooden fence posts and so on. We have had some big ones but none to my knowledge have gone out of control.

Another factor in planning the work is the bird-nesting season when our repertoire sometimes extends to other activities. One of the more unusual tasks was manoeuvring several vintage coaches stored on temporary tracks in the yard at Williton. After a day's work with the CBG, there's great satisfaction in seeing a job well done, with teamwork which allows everyone to work as hard and for as long as they wish.

In 1999 Ian Jonas asked for volunteers to manage the re-growth on any section of the line, typically half-a-mile, and he provided a set of instructions. Peter Gleed and I ("Gleed With Ritchie" - GWR!) chose a section between Bishops Lydeard and Crowcombe Heathfield between the bridges at Churchlands and Nethercott. This is relatively convenient as we travel from the south – we both live in Devon – and parking is available by both bridges. We can also appreciate engines working hard upgrade! Peter and I have enjoyed many days out to places of railway interest, so we are the "Boys Railway Appreciation Society (BRAS)." Thursday is our weekly Lengthmen day provided the weather is conducive. Our team has changed over the years and is currently Hugh Gloyn, Paul Murray, Phil Wall and myself as the Person in Charge. Paul does the strimming while the others use hand tools – bow saws, branch pruners, and heavy-duty shears. Our bonfires are planned to avoid smoke drifting into neighbouring properties or across roads. Engine smoke is OK as it clears quickly!

In recent years we have extended our section from Combe Florey bridge to the Nethercott footpath crossing. It now borders several houses, so we make a special effort to keep the neighbours on-side by giving special attention to "their" embankments. We know our efforts are appreciated when coffee or tea are offered by the 'KitKat Lady'.

Soon after the Norton site was acquired, David Holmes, Special Projects Officer, invited us to volunteer for an unusual and rather urgent task to find a sewer. We accepted so Peter, Hugh, Norman Barker and myself, with a sewer pipe/field plan as our guide, began to dig. It was a sunny day as we wielded our spades but the mood began to darken when we had reached the appropriate length and depth without success. Ian Jonas arrived with a water diviner, and the indications confirmed the presence of water, but after more digging without success, we accepted failure of our Great Sewer Pipe Hunt. The sewer was located on a later occasion by mechanical means, so it could then be protected. This event created another description for us: The Pensioners Rapid Response Team (PRRT).

We have also helped Derek Head with the renovation of the platelayer's hut at Combe Florey, and digging out the bank around the hut at Leigh Wood level crossing. Then on one occasion in the cutting at East Combe, the Dingle Brothers approached our work site as expected with their rail-mounted flail. We took shelter in the hut, which was battered for several minutes by flying debris including stones which, fortunately, didn't penetrate the reinforced glass window.

We've also worked at Watersmeet, East Combe/Tribble Bridge, and Combe Florey cutting back; Nornvis Bridge improving footpath visibility; Waterdene Farm improving visibility at the farm road crossing; Stogumber on re-growth behind the platform; Kingswood on re-growth; Doniford's pathway and hedges; Watchet on rubbish collection; and Dunster to improve level-crossing visibility at Sea Lane.

Occasionally we get involved with an unplanned non-railway activity. Peter and I were returning to our cars near Tribble Bridge at the end of the day when an articulated lorry emerged from the minor Lydeard St. Lawrence road on to the A358 towards Taunton. Confronted by the railway overbridge at Tribble, the driver stopped as he was uncertain whether the lorry would get through. We stopped the traffic in both directions since our high vis clothing gave an air of authority, and the driver eased the lorry forward. We judged that the clearance wouldn't be sufficient – we certainly didn't want a bridge strike – so the driver agreed there was no alternative but to reverse and return from where he came because in the other direction on the A358 is the similar-height railway overbridge at Combe Florey. We helped with the driver's reversing manoeuvre on the main road so he could return to the minor road. Traffic was held up for about 10 minutes but the situation could have been worse. All part of the service!

During the time David Holmes was in charge of Special Projects, our happy band of Lengthmen became Chopper Holmes's Active Pensioners' Squad (CHAPS). So whether we are BRAS, GWR, PRRT or CHAPS, we always look forward to our days when we do our bit to support the railway. If we're working near Churchlands Bridge, we can hear the birds singing and the whistle of an engine for the footpath crossing at Whiskey Trail, and we retreat to a safe distance from the line where we can be visible to the crew; the train appears; we acknowledge the engine's whistle; we wave to the locomotive crew and the passengers; the engine's exhaust fades away; then all is peaceful again. Time for a break or back to work? Let's have a break...and a KitKat. And if someone comments "We never see you doing any work" we don't mind at all as we have the satisfaction of seeing the results of our healthy exercise – a well-managed section of lineside with wild flowers in abundance.